Quantitative and Qualitative Games

This is Volume 58 in
MATHEMATICS IN SCIENCE AND ENGINEERING
A Series of monographs and textbooks
Edited by RICHARD BELLMAN, *University of Southern California*

A complete list of the books in this series appears at the end of this volume.

QUANTITATIVE AND QUALITATIVE GAMES

AUSTIN BLAQUIÈRE
FACULTY OF SCIENCES, PARIS
UNIVERSITY OF PARIS, FRANCE

FRANCOISE GÉRARD
FACULTY OF SCIENCES, PARIS
UNIVERSITY OF PARIS, FRANCE

GEORGE LEITMANN
DIVISION OF APPLIED MECHANICS
UNIVERSITY OF CALIFORNIA
BERKELEY, CALIFORNIA

ACADEMIC PRESS New York and London 1969

ACADEMIC PRESS, INC.
111 Fifth Avenue, New York, New York 10003

United Kingdom Edition published by
ACADEMIC PRESS, INC. (LONDON) LTD.
Berkeley Square House, London W1X 6BA

LIBRARY OF CONGRESS CATALOG CARD NUMBER: 74-84246
AMS 1968 SUBJECT CLASSIFICATION 9075

PRINTED IN THE UNITED STATES OF AMERICA

Preface

This book, based on lectures given by A. Blaquière at the Institute Henri Poincarè, is a record of joint research in the field of two-person games. Having arrived at a reasonably comprehensive theory for a wide class of such games, we present it here in the hope that it will serve as the basis for further research and for application in the physical, biological, and social sciences. Thus, this volume is addressed to those interested in game theory for its own sake, as well as to those concerned with conflict situations in technology, economics, psychology, indeed, in any dynamical system. With a view toward increasing the scope of the theory, the treatment proceeds from the general to the particular. Results are obtained for systems whose dynamics are defined by a rather large class of mappings. Further results are derived then by restricting the dynamics to those described by ordinary differential equations (differential games) and by difference equations (multistage games).

In Chapter 1, the notions of game and play are introduced, and the problem to be discussed is stated. Certain basic concepts, such as rules, objectives, and strategies, are defined.

Chapters 2–4 are devoted to quantitative two-person, zero-sum games with perfect information of the state called games of degree by Rufus Isaacs. These chapters are essentially extensions of the geometric theory of optimal processes developed earlier by A. Blaquière and G. Leitmann.† Differential as well as multi-stage games are treated.

Chapters 5–7 deal with qualitative two-player games, called games of

† See Refs. 1–6.

kind by Isaacs. In this type of game the players do not strive to minimize and maximize, respectively, a numerical payoff; rather they have conflicting aims in that they endeavor to terminate play on different target sets. In Chapter 5 additional geometric notions are introduced, such as the map of the game, and optimality for a player at a given state is defined. While the definitions and arguments are different from those employed in earlier chapters, the geometric results exhibit a strong similarity to the ones found in quantitative games. This similarity serves as a guide for the development of the theory. In Chapter 6, the theory is applied to differential games. The arguments used follow closely those employed in Chapter 3. While the framework of the theory of quantitative and qualitative games appears to be essentially the same, the similarities appear to be more formal than real. In Chapter 7, an attempt is made to establish a connection between these two types of games.

We believe that geometric notions are well suited for a discussion of dynamical system theory, such as that of the games treated in this book. Geometric notions are intuitively appealing and, at the same time, capable of providing a modicum of rigor. It is our hope that the reader will come to share this point of view and that he will not be discouraged by the many assumptions introduced in the development of the theory. The reader whose interest lies primarily in the application of the theory may elect to forego some of the details and accept many of the arguments on an intuitive basis. It has been our aim to state all assumptions explicitly rather than to leave it to the reader to discover them (or not). The complications which have necessitated many of these assumptions appear to be inherent in the theory. In any event, this book is but a brief introduction to a geometric theory of games, and many questions remain unanswered.

We should like to take this opportunity to thank the Office of Naval Research and the Delegation Generale à la Recherche Scientifique et Technique for supporting the research that has resulted in this book.

A. BLAQUIÈRE
F. GÉRARD
G. LEITMANN

Paris and Berkeley
September 1969

Contents

Notation and Terminology

Unless otherwise indicated mathematical symbols will be those in common use. Additional terminology will be introduced as needed.

Vectors will be denoted by lower case letters. Subscripts will be used to denote the components of a vector; superscripts will be used to distinguish vectors. $x = (x_1, x_2, \ldots x_n)$ is an element of E^n. No special notation is used to distinguish between $x \in E^n$ and its column form; that is, unless otherwise specified, all vectors will be column vectors. The row form of x is $x^T = (x_1\, x_2 \ldots x_n)$. The Euclidean length of a vector x will be denoted by $||x||$. The gradient of a function Φ: $x \rightarrow \Phi(x)$ at a point x^j will be denoted by grad $\Phi(x^j)$. The transpose of a matrix M will be denoted by M^T. The topological closure of a set Ω in a topological space will be denoted by $\bar{\Omega}$, and the interior of Ω by $\mathring{\Omega}$.

The term domain will mean an open connected set of the appropriate space. A scalar-valued function defined on a domain will be said to be of class C^k if the function and its first k partial derivatives are continuous on the domain. A vector-valued function will be said to be of class C^k if each of its components is of class C^k.

The notation $\forall\ v\ \exists\ u$ will stand for:

"for all v there exists u (*which depends on* v)"

and $\exists\ v\ \forall\ u$ will stand for:

"there exists u (*independent of* v) for all v"

We shall write *Definition* whenever we define a new concept, and not for the specification of concepts introduced previously.

We shall write *Assumption* whenever we introduce a new assumption of basic importance for the development of the theory.

Quantitative
and Qualitative Games

1 *Games and Plays*

1.1 PROBLEM STATEMENT AND ASSUMPTIONS

We shall be interested in the behavior of a set of "persons," called the *players*, each of whom strives to modify the state of a system in a most efficacious manner according to his own criterion. If there is only one player in the set, the theory reduces to the one of optimal control. In the following discussion we shall consider only the case of two players J_P and J_E whose interests are conflicting or at least different.

We shall assume that the state of the system or, as we shall say, the state of the *game*, at any instant of time can be characterized by the values of a finite number of real variables $x_1, x_2, \ldots, x_n$, and that these values are known to both players. Hence we may think of the state as a point in an n-dimensional Euclidean space E^n, termed the *state space* of the game. In state space we select a rectangular coordinate system so that we may define a point by the *state vector* $x = (x_1, x_2, \ldots, x_n)$.

We shall assume that the evolution in time of the state of the game can be influenced at all times by the two players. For example, as we shall see later, the behavior of the state may be described by a set of differential equations

$$dx_\nu/dt = f_\nu(x_1, x_2, \ldots, x_n, u_1, u_2, \ldots, u_r, v_1, v_2, \ldots, v_s), \qquad \nu = 1, 2, \ldots, n \quad (1.1)$$

where $u_1, u_2, \ldots, u_r$ and $v_1, v_2, \ldots, v_s$ are *control variables* which define vectors $u = (u_1, u_2, \ldots, u_r)$ and $v = (v_1, v_2, \ldots, v_s)$ in r-dimensional and s-dimensional Euclidean spaces E^r and E^s, respectively.

Players J_P and J_E make their decisions through choosing the values of control variables u and v, respectively, at each instant of time. Later we shall discuss the case in which these choices are governed by functions of x, p, and e, which J_P and J_E select from two prescribed sets of functions.

1.2 RULES OF THE GAME

We shall assume that x belongs to the fixed subset G of E^n.

Let there be given two prescribed sets R_P and R_E whose members will be denoted by r_P and r_E, respectively.

Let T be the collection of all subsets of $[0, \infty)$ containing 0.

Let F be the set of all functions $\mathbf{x}\colon\ T_\mathbf{x} \to G$ for all $T_\mathbf{x} \in T$. Hereafter we shall use simply

$$\mathbf{x}\colon\quad \tau \to x = \mathbf{x}(\tau), \qquad \tau \in T_\mathbf{x}$$

to indicate a function $\mathbf{x} \in F$.

Let $\mathscr{F}$ be the collection of all nonempty subsets of F, and let

$$\gamma\colon\quad G \times R_P \times R_E \to \mathscr{F}$$

be a mapping that associates with each point $x^i \in G$ and with each pair $(r_P, r_E) \in R_P \times R_E$ an element $\gamma(x^i, r_P, r_E)$ of $\mathscr{F}$, such that

$$\mathbf{x}(0) = x^i \qquad \forall \mathbf{x} \in \gamma(x^i, r_P, r_E)$$

Finally, let there be given two sets of points in G, namely θ_P and θ_E. We shall call these sets the *targets* of J_P and J_E, respectively.

Definition 1.1. A *game* is defined by its *rules* and its *objectives*.

The rules of the game consist of

(i) the set G

(ii) the sets R_P and R_E

(iii) the mapping γ.

The objectives will be defined later.

Definition 1.2. Members r_P of R_P and r_E of R_E are *strategies* of players J_P and J_E, respectively. Members (r_P, r_E) of $R_P \times R_E$ are *strategy pairs*.

EXAMPLE 1.1. Consider the following rules of a game:

Let G be a domain of E^n. Let there be given the differential equation

$$dx/d\tau = f(x, u, v) \tag{1.2}$$

where $u = (u_1, u_2, \ldots, u_r)$, $v = (v_1, v_2, \ldots, v_s)$, $f = (f_1, f_2, \ldots, f_n)$.

Let the choices of control variables u and v be governed by functions $p\colon\ x \to p(x)$, $e\colon\ x \to e(x)$, $x \in G$, respectively; that is, $r_P = p$ and $r_E = e$ being chosen from prescribed sets of functions R_P and R_E, respectively, let Eq. (1.2) be replaced by

$$dx/d\tau = f(x, p(x), e(x)) \tag{1.3}$$

Equation (1.3) is said to have a solution if there exists a function $\mathbf{x}\colon\ \tau \to \mathbf{x}(\tau)$ defined and continuous *on an interval* $\Delta\tau$, such that $\mathbf{x}(\tau) \in G$ $\forall \tau \in \Delta\tau$, and

$$d\mathbf{x}(\tau)/d\tau = f(\mathbf{x}(\tau), p(\mathbf{x}(\tau)), e(\mathbf{x}(\tau)))$$

is satisfied almost everywhere in $\Delta\tau$.

Now, for each initial point $x^i \in G$ there may exist a solution of Eq. (1.3), $\mathbf{x}\colon\ \tau \to \mathbf{x}(\tau)$, in† $[0, \infty)$ such that $\mathbf{x}(0) = x^i$. Let $[0, \tau_s)$ be its interval of definition.

We shall call $\mathbf{x}$ a *maximal solution* if $\mathbf{x}(\tau)$ is not defined for $\tau = \tau_s$, or if $\lim_{\tau\to\tau_s, \tau<\tau_s} \mathbf{x}(\tau)$ is defined and does not belong to G, or if $\lim_{\tau\to\tau_s,\tau<\tau_s} \mathbf{x}(\tau)$ is defined and belongs to G and Eq. (1.3) has no solution $\hat{\mathbf{x}}\colon\ \tau \to \hat{\mathbf{x}}(\tau)$ in $[0, \infty)$ that satisfies

$$\hat{\mathbf{x}}(0) = x^s, \qquad x^s \triangleq \lim_{\substack{\tau\to\tau_s \\ \tau<\tau_s}} \mathbf{x}(\tau).$$

Then τ_s will be called the *extreme future value* of τ for the maximal solution and denoted by τ^+.

Next let us define $\gamma(x^i, r_P, r_E)$ for all $x^i \in G$, all $r_P \in R_P$ and all $r_E \in R_E$, as follows:

(i) If Eq. (1.3) has no solution $\mathbf{x}\colon\ \tau \to \mathbf{x}(\tau)$ in $[0, \infty)$ such that $\mathbf{x}(0) = x^i$, then we let $\gamma(x^i, r_P, r_E)$ be the point x^i itself, that is the function $\mathbf{x}$ whose set of definition is $\tau = 0$ such that $\mathbf{x}(0) = x^i$;

(ii) otherwise we let $\gamma(x^i, r_P, r_E)$ be the set of all maximal solutions $\mathbf{x}\colon\ \tau \to \mathbf{x}(\tau)$ of Eq. (1.3) in $[0, \infty)$, such that $\mathbf{x}(0) = x^i$.

Hence γ is defined. The rules of the game consist of G, the sets of functions p and e, and mapping γ.

1.3 TRAJECTORIES AND PATHS

Definition 1.3. Given the rules, for given strategies r_P and r_E chosen by players J_P and J_E from sets R_P and R_E, respectively, and for a given state

† That is, on some interval $\Delta\tau \in [0, \infty)$.

$x^i \in G$ called the *initial state*, a curve $\Gamma_{\mathbf{x}}(x^i, r_P, r_E)$ in G defined by a function $\mathbf{x} \in \gamma(x^i, r_P, r_E)$ will be called a *describing curve*.

Of course, there may be more than one describing curve for given initial state x^i and given strategies r_P and r_E.

We shall also admit the possibility that $\Gamma_{\mathbf{x}}(x^i, r_P, r_E)$ is a single point.

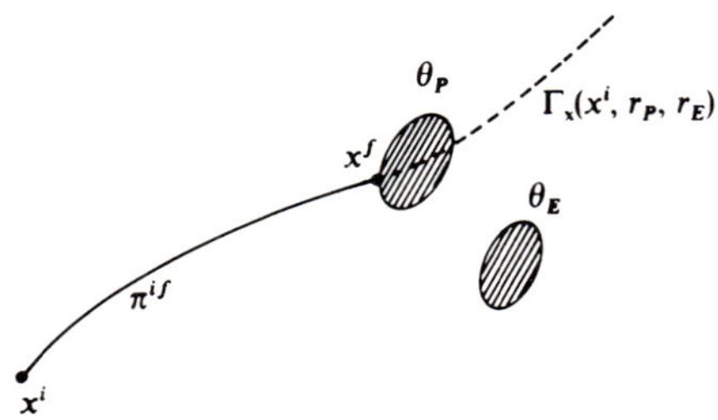

FIG. 1.1. Path π^{if} in state space, terminating play.

Definition 1.4. Given any describing curve $\Gamma_{\mathbf{x}}(x^i, r_P, r_E)$ which emanates from x^i and is defined by function $\mathbf{x}$ on $T_{\mathbf{x}}$, and given $\tau_j \in T_{\mathbf{x}}$, the image of $[0, \tau_j]$ under the graph $\{(\tau, x): \; x = \mathbf{x}(\tau)\}$ will be called a *trajectory* and denoted by Γ^{ij}. Its *initial point* is $x^i = \mathbf{x}(0)$, its *end point* is $x^j = \mathbf{x}(\tau_j)$ and its *supporting curve* is $\Gamma_{\mathbf{x}}(x^i, r_P, r_E)$.

We shall say that Γ^{ij} is generated by (r_P, r_E). We shall also admit the possibility that $\tau_j = 0$.

Definition 1.5. A trajectory Γ^{ij} defined by function $\mathbf{x}$ on $[0, \tau_j] \cap T_{\mathbf{x}}$, $\tau^j \neq 0$, will be called a *path* and denoted by π^{ij}, if

$$\mathbf{x}(\tau) \notin \theta_P \cup \theta_E \qquad \forall \tau \in [0, \tau_j) \cap T_{\mathbf{x}} \tag{1.4}$$

It follows from that definition that no point of a path, with the possible exception of its end point, belongs to θ_P or θ_E.

If $\tau_j = 0$, Γ^{ij} is said to be a *null path*, and will be denoted by π^0.

1.4 PLAY, TERMINATING PLAY

Definition 1.6. A *play* is the evolution of state x, described by $\mathbf{x}$: $\tau \to x = \mathbf{x}(\tau)$, $\tau \in [0, \tau_j]$ along a path π^{ij}.

A *terminating play* is a play whose end state $x^f = \mathbf{x}(\tau_f)$ belongs to $\theta_P \cup \theta_E$. It *terminates* at $\tau = \tau_f$. The corresponding path will be called a *terminating path.*

We shall say that a strategy pair (r_P, r_E) generates a terminating play from initial state x^i if there exists a function $\mathbf{x} \in \gamma(x^i, r_P, r_E)$ that defines a terminating path.

1.5 PLAYABLE STRATEGY PAIRS

In the following discussion, we shall consider two cases, namely,

Case (a): neither θ_P nor θ_E is empty;

and

Case (b): θ_P or θ_E is empty, but not both.

For instance, we shall suppose that $\theta_P \neq \varnothing$ and $\theta_E = \varnothing$.

Definition 1.7. *In Case (a) we shall say that (r_P, r_E) is a weakly playable*† *strategy pair for player J_P at point x^i, if it generates a terminating play from initial state x^i, and if the corresponding end state x^f belongs to θ_P; and we shall say that (r_P, r_E) is a weakly playable*† *strategy pair for player J_E at point x^i, if it generates a terminating play from initial state x^i, and if the corresponding end state x^f belongs to θ_E.*

In Case (b) the definition of a weakly playable† *strategy pair for J_P at point x^i will be the same as in Case (a); and we shall say that (r_P, r_E) is a weakly playable*† *strategy pair for J_E at point x^i, if it is not playable for J_P at that point.*

We shall let $\mathscr{T}_P(x^i)$ and $\mathscr{T}_E(x^i)$ denote the sets of all weakly playable strategy pairs for J_P and J_E, respectively, at point x^i.

It follows from Definition 1.7 that in *Case (a)*

$$(r_P, r_E) \in \mathscr{T}_P(x^i) \Rightarrow \exists \mathbf{x} \in \gamma(x^i, r_P, r_E)$$

such that

$$\Gamma_{\mathbf{x}}(x^i, r_P, r_E) \cap \theta_P \neq \varnothing$$

$$(r_P, r_E) \in \mathscr{T}_E(x^i) \Rightarrow \exists \mathbf{x} \in \gamma(x^i, r_P, r_E)$$

† Or, more simply, "playable."

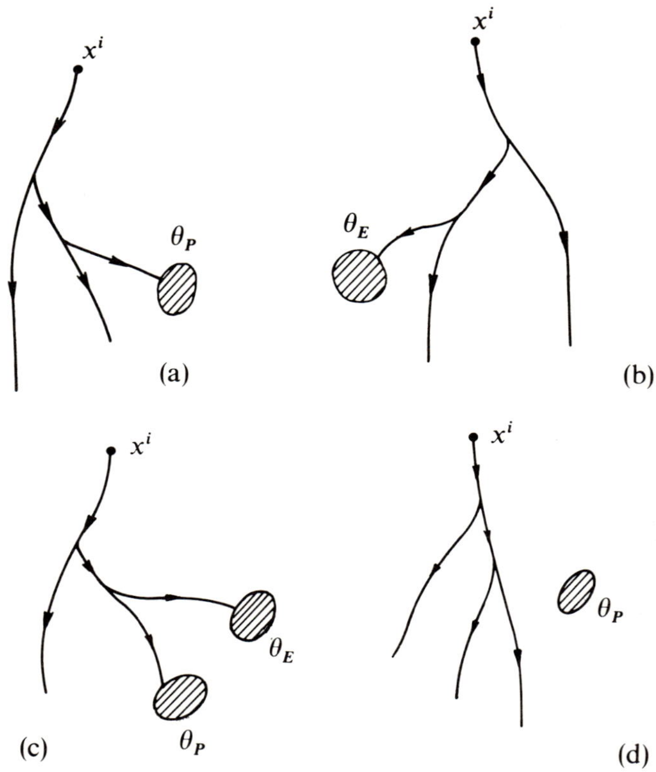

FIG. 1.2. Paths generated by a playable strategy pair for (a) J_P at x^i; (b) J_E at x^i; (c) both J_P and J_E at x^i; (d) J_E at x^i in a single target game.

such that
$$\Gamma_{\mathbf{x}}(x^i, r_P, r_E) \cap \theta_E \neq \varnothing$$

and in *Case (b)*,

$$(r_P, r_E) \in \mathscr{T}_P(x) \Leftrightarrow \exists \mathbf{x} \in \gamma(x^i, r_P, r_E)$$

such that
$$\Gamma_{\mathbf{x}}(x^i, r_P, r_E) \cap \theta_P \neq \varnothing$$

$$(r_P, r_E) \in \mathscr{T}_E(x^i) \Leftrightarrow \forall \mathbf{x} \in \gamma(x^i, r_P, r_E)$$

$$\Gamma_{\mathbf{x}}(x^i, r_P, r_E) \cap \theta_P = \varnothing$$

Definition 1.8. *In Case* (*a*) we shall say that a strategy pair (r_P, r_E) is *strongly playable for J_P at point x^i*, if

(i) it is not playable for J_E at x^i; and

(ii) $\Gamma_{\mathbf{x}}(x^i, r_P, r_E) \cap \theta_P \neq \varnothing \;\; \forall \mathbf{x} \in \gamma(x^i, r_P, r_E)$

Likewise a strategy pair (r_P, r_E) will be said to be *strongly playable for J_E at point x^i*, if

(i)′ it is not playable for J_P at x^i; and

(ii)′ $\Gamma_{\mathbf{x}}(x^i, r_P, r_E) \cap \theta_E \neq \varnothing \;\; \forall \mathbf{x} \in \gamma(x^i, r_P, r_E)$

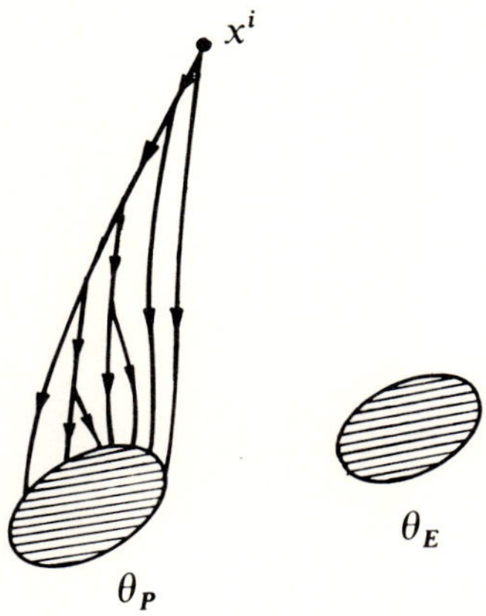

FIG. 1.3. Paths generated by a strongly playable strategy pair for J_P at point x^i.

In Case (*b*) a strategy pair (r_P, r_E) will be said to be strongly playable for J_P at point x^i, if

$$\Gamma_{\mathbf{x}}(x^i, r_P, r_E) \cap \theta_P \neq \varnothing \qquad \forall \mathbf{x} \in \gamma(x^i, r_P, r_E)$$

There is no difference between the concepts of weak playability and strong playability for J_E.

Of course, in Case (b), playability for J_E at point x^i means that *no* describing curve $\Gamma_{\mathbf{x}}(x^i, r_P, r_E)$ intersects θ_P and so this concept is sufficiently strong by itself.

Note that, in every case, strong playability implies weak playability.

We shall let $\mathscr{T}_P{}^*(x^i)$ and $\mathscr{T}_E{}^*(x^i)$ denote the sets of all strongly playable pairs for J_P and J_E, respectively, at point x^i.

1.6 JOINING OF TRAJECTORIES AND DESCRIBING CURVES

Definition 1.9. Let Γ^{ij} be a trajectory emanating from x^i and ending at x^j; and let $\Gamma_{\mathbf{x}}(x^j, \hat{r}_P, \hat{r}_E)$ be a describing curve emanating from x^j. We shall say that Γ^{ij} and $\Gamma_{\mathbf{x}}(x^j, \hat{r}_P, \hat{r}_E)$ can be *joined* if

$$\Gamma^{ij} \cup \Gamma_{\mathbf{x}}(x^j, \hat{r}_P, \hat{r}_E)$$

is a describing curve emanating from x^i, that is if there exist a strategy

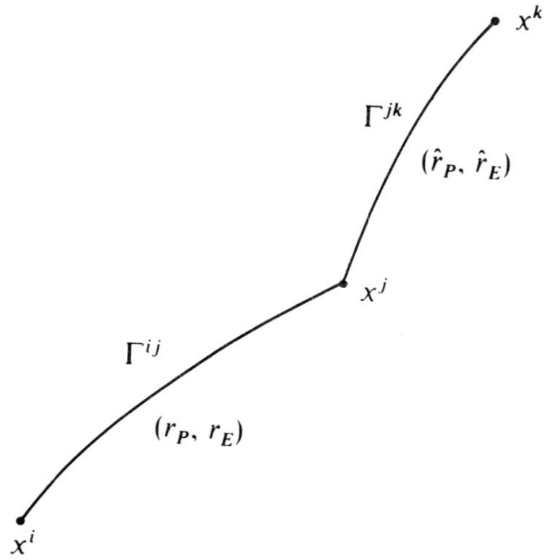

FIG. 1.4. Joining of trajectories. $\Gamma^{ik} = \Gamma^{ij} \cup \Gamma^{jk}$ is a trajectory generated by (ρ_P, ρ_E).

pair $(\rho_P, \rho_E) \in R_P \times R_E$ and a function $\xi \in \gamma(x^i, \rho_P, \rho_E)$, such that

$$\Gamma_\xi(x^i, \rho_P, \rho_E) = \Gamma^{ij} \cup \Gamma_{\mathbf{x}}(x^j, \hat{r}_P, \hat{r}_E)$$

Definition 1.10. We shall say that trajectories Γ^{ij} and Γ^{jk} can be *joined* if $\Gamma^{ij} \cup \Gamma^{jk}$ is a trajectory Γ^{ik}.

One can readily verify that if two paths† π^{ij} and π^{jk} can be joined, then $\pi^{ij} \cup \pi^{jk}$ is a path π^{ik}.

Unless otherwise indicated we shall adopt

† Recall that a path is a trajectory. However, the converse need not be true.

Assumption 1.1. Every trajectory Γ^{ij}, generated by (r_P, r_E) and ending at x^j, can be joined with describing curve $\Gamma_{\mathbf{x}}(x^j, \hat{r}_P, \hat{r}_E)$, no matter what $\hat{r}_P \in R_P$, $\hat{r}_E \in R_E$ and $\mathbf{x} \in \gamma(x^j, \hat{r}_P, \hat{r}_E)$. Furthermore, if $\hat{r}_P = r_P$, there exists a ρ_P that is equal to r_P; and if $\hat{r}_E = r_E$, there exists a ρ_E that is equal to r_E.

As a direct consequence of Assumption 1.1, all trajectories Γ^{ij} and Γ^{jk} can be joined; and

(i) if Γ^{ij} and Γ^{jk} are generated by (r_P, r_E) and $(r_P, \hat{r}_E)$, respectively, then $\Gamma^{ij} \cup \Gamma^{jk}$ is generated by a strategy pair (r_P, ρ_E);

(ii) if Γ^{ij} and Γ^{jk} are generated by (r_P, r_E) and $(\hat{r}_P, r_E)$, respectively, then $\Gamma^{ij} \cup \Gamma^{jk}$ is generated by a strategy pair (ρ_P, r_E).

1.7 OBJECTIVES OF THE GAME, QUALITATIVE AND QUANTITATIVE GAMES

Definition 1.11. The *objectives* of the game are the targets, θ_P and θ_E in Case (a), and θ_P or θ_E in Case (b), and two prescriptions, one for player J_P and the other for player J_E. These prescriptions depend on the type of game considered. (They are given below for the two cases studied in this book.)

1. A *qualitative game* (or game of kind†) is a game in which the criterion is qualitative. In Case (a), for any given $x^i \in G$, the prescription for player J_P is to use a strategy $r_P \in R_P$ such that (r_P, r_E) generates a terminating play with end state $x^f \in \theta_P$, no matter what the strategy $r_E \in R_E$ of J_E. We shall say that his prescription is to transfer the state to θ_P. The prescription for player J_E is to transfer the state to θ_E. Of course, one or the other player, or both, may be unable to control the play in such a manner that the condition which has been prescribed for him is satisfied.

Definition 1.12. In a qualitative game, we shall say that a player is a *winner* if he has transferred the state of the game to his own target, despite the efforts of his opponent.

† According to the terminology of Isaacs.

Both players can win if the intersection of θ_P and θ_E is not empty.

In Case (b), $\theta_P \neq \varnothing$, $\theta_E = \varnothing$ (or $\theta_E \neq \varnothing$, $\theta_P = \varnothing$) the prescription for player J_P (or J_E) is to transfer the state to θ_P (or θ_E), and the prescription for his opponent, J_E (or J_P), is to prevent this event.

If J_P (or J_E) transfers the state of the game to his target despite the efforts of his opponent, he is the winner. According to Definition 1.12, the player who wishes to prevent the state from reaching the target cannot win since he has no target of his own.

2. A *quantitative game* (or game of degree†) is a game in which (i) there is given a single target‡ $\theta = \theta_P = \theta_E$; and (ii) there is given a functional which assigns a unique real number to each transfer effected by a strategy pair along a path. Such a functional is called a *performance index*, and the number which it assigns to a transfer is called the value of the functional, or the *cost*.

In zero-sum, two player games, the prescription for one of the players is to maximize the cost of transfer from an initial state to the target, while the prescription for the other is to minimize it. In such games there is no winner, since the outcome of the game is a compromise between the players.

† According to the terminology of Isaacs.

‡ This case is Case (a) with $\theta_P = \theta_E$.

II *Some Geometric Aspects of Quantitative Games*

2.1 TRANSFER OF STATE AND PERFORMANCE INDEX

As we have seen in Chapter 1, given an initial state x^i in G, the rules of the game, the strategies r_P and r_E chosen by players J_P and J_E from prescribed sets R_P and R_E, respectively, and a function $\mathbf{x} \in \gamma(x^i, r_P, r_E)$, the state of the game moves along describing curve $\Gamma_{\mathbf{x}}(x^i, r_P, r_E)$.

Here we shall consider the paths which are generated by all strategy pairs. In particular, we shall be concerned with transferring the state along a path from an initial state x^i to any one in prescribed set of states θ in G. In general, some paths will correspond to a transfer of the state from x^i to some terminal point x^f in θ, while others will correspond to a transfer from x^i to a terminal point which does not belong to θ.

Definition 2.1. Next let us adopt a functional which assigns a unique real number to each transfer effected by a strategy pair along a path. Such a functional will be called a *performance index*, and the number which it assigns to a transfer from initial point x^i to end point x^s, along a path π^{is} in G, will be called the *cost* of the transfer.

We shall admit a performance index such that the cost of transfer from x^i to x^s along path π^{is} depends on the generating strategies r_P and r_E. We shall denote the cost of such a transfer by $\mathscr{V}(x^i, x^s;\ r_P, r_E, \pi^{is})$.

Now we introduce

Assumption 2.1. The cost which is associated with a null path π^0 is zero; that is

$$\mathscr{V}(x^i, x^i;\ r_P, r_E, \pi^0) = 0 \qquad \forall r_P \in R_P \qquad \forall r_E \in R_E \tag{2.1}$$

2.2 OPTIMALITY

It will be convenient to consider a quantitative game as a game with two targets θ_P and θ_E, for which

$$\theta_P = \theta_E = \theta$$

Then, for any initial state x^i, a strategy pair (r_P, r_E) which is playable for J_P at point x^i is also playable for J_E at that point. Accordingly, in that case, we shall say that (r_P, r_E) is a *playable strategy pair at point* x^i, without specifying the player who is concerned, and we shall let $\mathscr{T}(x^i)$ denote the set of all playable pairs (r_P, r_E) at point x^i.

We shall let $V(x^i, \theta;\ r_P, r_E)$ denote the cost of transfer from x^i to θ along a path π^{if} generated by $(r_P, r_E) \in \mathscr{T}(x^i)$. Since, for given x^i and $(r_P, r_E) \in \mathscr{T}(x^i)$, π^{if} need not be unique, $V(x^i, \theta;\ r_P, r_E)$ need not be unique. Along π^{if} we have

$$V(x^i, \theta;\ r_P, r_E) = \mathscr{V}(x^i, x^f;\ r_P, r_E, \pi^{if}) \tag{2.2}$$

Definition 2.2. Let there be given a fixed set $X \subseteq G$. We shall say that a strategy pair (r_P^*, r_E^*) is *optimal on* X, if

(i) it is playable at all initial points $x^i \in X$; and

(ii) $V(x^i, \theta;\ r_P^*, r_E^*)$ is defined for all $x^i \in X$; that is, if there is more than one path starting at point x^i, generated by (r_P^*, r_E^*) and terminating on θ, then all the values of $V(x^i, \theta;\ r_P^*, r_E^*)$ are equal;

(iii) the saddle-point condition

$$V(x^i, \theta;\ r_P^*, r_E) \leqslant V(x^i, \theta;\ r_P^*, r_E^*) \leqslant V(x^i, \theta;\ r_P, r_E^*) \tag{2.3}$$

holds for all $x^i \in X$ and for all strategy pairs $(r_P^*, r_E) \in \mathscr{T}(x^i)$ and $(r_P, r_E^*) \in \mathscr{T}(x^i)$.

We emphasize that (2.3) must hold for all values of $V(x^i, \theta;\ r_P^*, r_E)$ and all values of $V(x^i, \theta;\ r_P, r_E^*)$.

Definition 2.3. Since $V(x^i, \theta;\ r_P^*, r_E^*)$ is defined for all $x^i \in X$, $V^*\colon x^i \to V^*(x^i)$, where

$$V^*(x^i) \triangleq V(x^i, \theta;\ r_P^*, r_E^*) \tag{2.4}$$

is a (single-valued) function on X. We shall call $V^*(x^i)$ the *Value of the game* for play starting at x^i.

If (i)–(iii) are satisfied for all $x^i \in X$, then, by Assumption 2.1, (i)–(iii) are satisfied for all $x^i \in X \cup \theta$; and, of course, for $x^i \in \theta$ we have

$$V^*(x^i) = 0.$$

Let $X^* \triangleq X \cup \theta$.

2.3 GAME AND ISOVALUE SURFACES

Let us introduce another variable, x_0, and consider an $(n + 1)$-dimensional Euclidean space E^{n+1} of points x, where $\mathsf{x} = (x_0, x) = (x_0, x_1, \ldots, x_n)$ is the vector which defines a point relative to a rectangular coordinate system in E^{n+1}, the *augmented state space.*

Definition 2.4. Since V^*: $x^i \to V^*(x^i)$ is defined on X^*, we can define a *game surface* $\Sigma(C)$ in $\mathscr{X}^* \triangleq X^* \times \{x_0\}$ by

$$\Sigma(C) \triangleq \{\mathsf{x}: \quad x_0 + V^*(x) = C\} \tag{2.5}$$

where C is a constant parameter.

The intersection of $\Sigma(C)$ with X^* will be called an *isovalue surface*†

$$S(C) \triangleq \{x: \quad V^*(x) = C\} \tag{2.6}$$

Surface $S(C)$ is the locus of all initial states for which the Value of the game is the same, namely C; hence the name isovalue surface.

As the value of parameter C is varied, Eqs. (2.5) and (2.6) define one-parameter families of surfaces $\{\Sigma(C)\}$ and $\{S(C)\}$.

For each given C, Eq. (2.5) defines a single-sheeted surface in $\mathscr{X}^*$; that is, $\Sigma(C)$ is a set of points which are in one-to-one correspondence with the points of X^*.

Consider now two game surfaces $\Sigma(C_1)$ and $\Sigma(C_2)$, corresponding to two different parameter values C_1 and C_2, respectively. Let x_{01} and x_{02} denote the values of x_0 on $\Sigma(C_1)$ and $\Sigma(C_2)$, respectively, for the same state. Then it follows from (2.5) that

$$x_{01} - x_{02} = C_1 - C_2 \qquad \forall x \in X^*$$

Consequently, the members of the one-parameter family $\{\Sigma(C)\}$ may be deduced from one another by translation parallel to the x_0-axis. Furthermore, these surfaces are ordered along the x_0-axis in the same way as the parameter value C; that is, the "higher" a surface in the family the

† Note that $S(C)$ may be an empty set.

greater is the parameter value C. Thus one and only one game surface passes through a given point x in $\mathscr{X}^*$.

Now let us introduce some nomenclature.

Definition 2.5. A given game surface $\Sigma(C)$ separates the set $\mathscr{X}^*$ into two disjoint sets. We shall denote these sets by $A/\Sigma(C)$ ["above" $\Sigma(C)$] and $B/\Sigma(C)$ ["below $\Sigma(C)$"], respectively. For a game surface $\Sigma(C)$

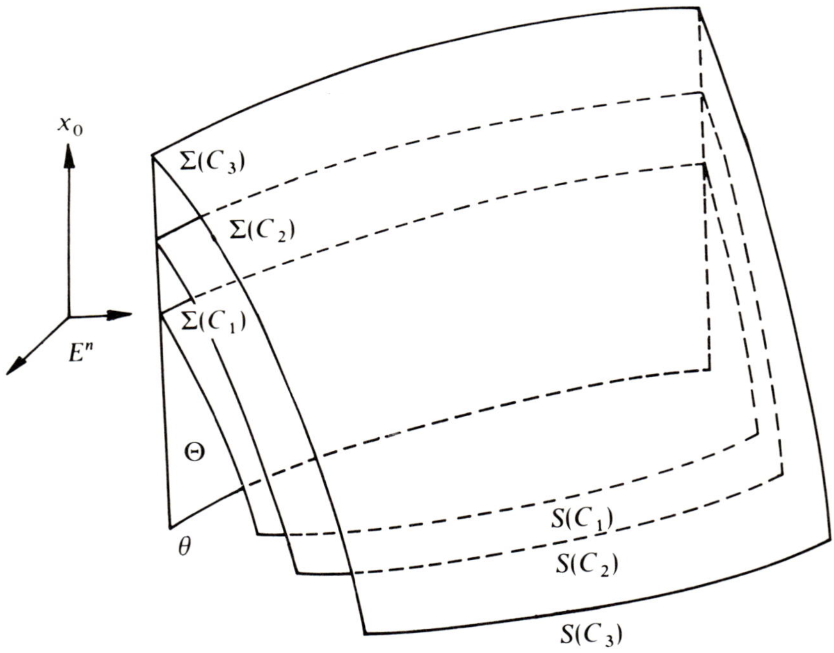

FIG. 2.1. Game and isovalue surfaces.

corresponding to parameter value C, we have

$$A/\Sigma(C) \triangleq \{x: \quad x_0 > C - V^*(x)\} \tag{2.7}$$

$$B/\Sigma(C) \triangleq \{x: \quad x_0 < C - V^*(x)\} \tag{2.8}$$

A point $x \in A/\Sigma(C)$ will be called an A-point relative to $\Sigma(C)$, and a point $x \in B/\Sigma(C)$ a B-point relative to $\Sigma(C)$.

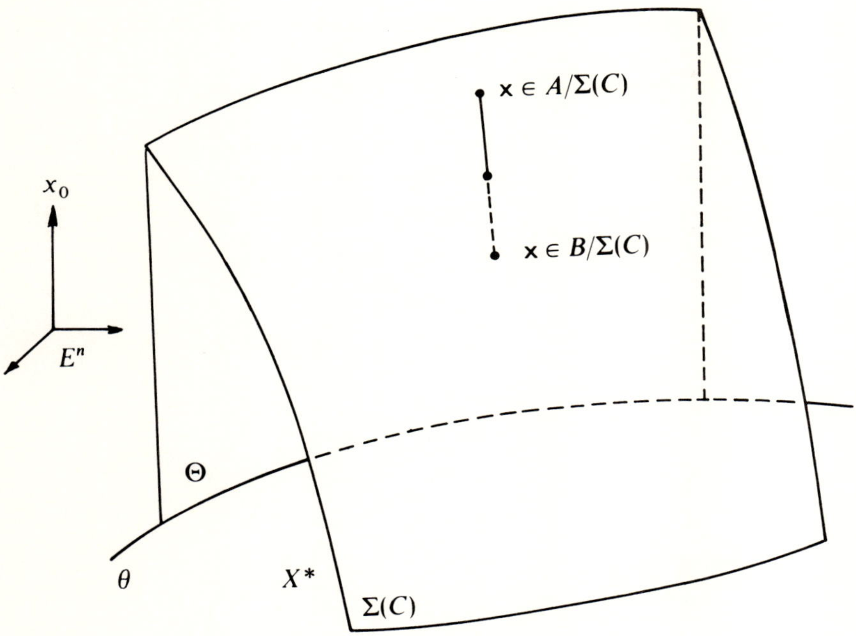

FIG. 2.2. A-point and B-point relative to $\Sigma(C)$.

Hence, the equation of the game surface through a point $\mathbf{x}^i \in \mathscr{X}^*$ is

$$x_0 + V^*(x) = x_0{}^i + V^*(x^i) = C \tag{2.9}$$

and

$$A/\Sigma(C) = \{\mathbf{x}: \quad x_0 - x_0{}^i > V^*(x^i) - V^*(x)\} \tag{2.10}$$

$$B/\Sigma(C) = \{\mathbf{x}: \quad x_0 - x_0{}^i < V^*(x^i) - V^*(x)\} \tag{2.11}$$

2.4 PATHS IN AUGMENTED STATE SPACE

Definition 2.6. Paths $\Pi^{is}(C)$ in $\mathscr{G} \triangleq G \times \{x_0\}$ are defined by

$$\Pi^{is}(C) \triangleq \{\mathbf{x}^k: \quad x_0{}^k + \mathscr{V}(x^k, x^s;\ r_P, r_E, \pi^{ks}) = C,\ \pi^{ks} \subset \pi^{is}\} \tag{2.12}$$

where π^{is} is a path in G from x^i to x^s generated by strategies $r_P \in R_P$ and $r_E \in R_E$, and C is a constant parameter.

Thus, path π^{is} is the projection on G of paths $\Pi^{is}(C)$ in $\mathscr{G}$.

Let $\Theta \triangleq \theta \times \{x_0\}$ be the target in augmented state space. One can

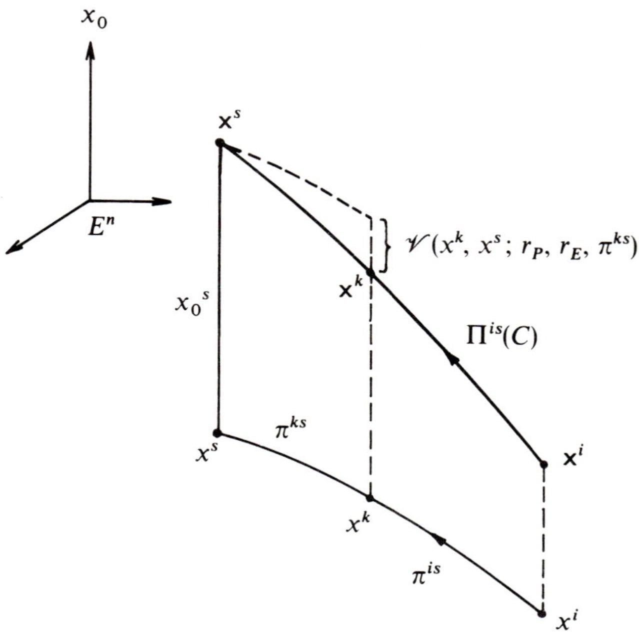

FIG. 2.3. Path in augmented state space.

easily verify that the above definition of a path in $\mathscr{G}$ agrees with the general definition of a path given in Section 1.3.

By varying the value of parameter C in (2.12) one generates a one-parameter family of paths $\{\Pi^{is}(C)\}$. This family belongs to an x_0-cylindrical surface whose intersection with G is path π^{is}.

Definition 2.7. Three families of paths emanating from points in $\mathscr{X}^*$, $\{\Pi_P^{ij}(C)\}$, $\{\Pi_E^{lm}(C)\}$, and $\{\Pi_{PE}^{nq}(C)\}$, are defined by

$$\Pi_P^{ij}(C) \triangleq \{\mathbf{x}^k\colon\ x_0{}^k + \mathscr{V}(x^k, x^j; r_P{}^*, r_E, \pi_P^{kj}) = C,\ \pi_P^{kj} \subset \pi_P^{ij}\} \tag{2.13}$$

$$\Pi_E^{lm}(C) \triangleq \{\mathbf{x}^k\colon\ x_0{}^k + \mathscr{V}(x^k, x^m; r_P, r_E{}^*, \pi_E^{km}) = C,\ \pi_E^{km} \subset \pi_E^{lm}\} \tag{2.14}$$

$$\Pi_{PE}^{nq}(C) \triangleq \{\mathbf{x}^k\colon x_0{}^k + \mathscr{V}(x^k, x^q; r_P{}^*, r_E{}^*, \pi_{PE}^{kq}) = C,\ \pi_{PE}^{kq} \subset \pi_{PE}^{nq}\} \tag{2.15}$$

Paths π_P^{ij}, π_E^{lm} and π_{PE}^{nq} are paths in G, emanating from points x^i, x^l, and x^n of X^*, generated by the strategy pairs (r_P^*, r_E), (r_P, r_E^*) and (r_P^*, r_E^*) and ending at points x^j, x^m and x^q, respectively.

Definition 2.8. Since (r_P^*, r_E^*) is playable at all points $x^n \in X^*$, there exists a path $\Pi_{PE}^{nf}(C)$ which reaches Θ at point x^f. It is called an *optimal path.*

Hereafter we shall denote an optimal path by $\Pi^*(C)$.

Definition 2.9. If (r_P^*, r_E) and (r_P, r_E^*) are playable strategy pairs at points x^i and x^l, respectively, there also exist paths $\Pi_P^{if}(C)$ and $\Pi_E^{lf}(C)$ which reach Θ.† They are called *P-optimal* and *E-optimal* paths, respectively.

Hereafter we shall write simply $\Pi_P(C)$ and $\Pi_E(C)$ to indicate these paths.

2.5 JOINING OF PATHS

Let π^{ij} be a path emanating from x^i, generated by (r_P, r_E), and ending at x^j; and let π^{js} be a path emanating from x^j, generated by $(\hat{r}_P, \hat{r}_E)$, and ending at x^s.

In addition to Assumption 1.1, according to which $\pi^{ij} \cup \pi^{js}$ is a path π^{is} generated by (ρ_P, ρ_E), we shall introduce

Assumption 2.2

$$\mathscr{V}(x^i, x^j;\ r_P, r_E, \pi^{ij}) + \mathscr{V}(x^j, x^s;\ \hat{r}_P, \hat{r}_E, \pi^{js}) \\ = \mathscr{V}(x^i, x^s;\ \rho_P, \rho_E, \pi^{is}) \qquad (2.16)$$

We shall call this assumption an *additivity property of the cost.*

2.6 SOME PROPERTIES OF GAME SURFACES

In preparation for a fundamental theorem, let us now prove

Lemma 2.1. *No point of a path $\Pi_P^{is}(C')$ which emanates from x^i is an A-point relative to the game surface through x^i.*

† Of course, superscript f means that the terminal point of the path belongs to Θ. However the terminal point need not be the same for paths $\Pi_P^{if}(C)$, $\Pi_E^{lf}(C)$, and $\Pi_{PE}^{nf}(C)$.

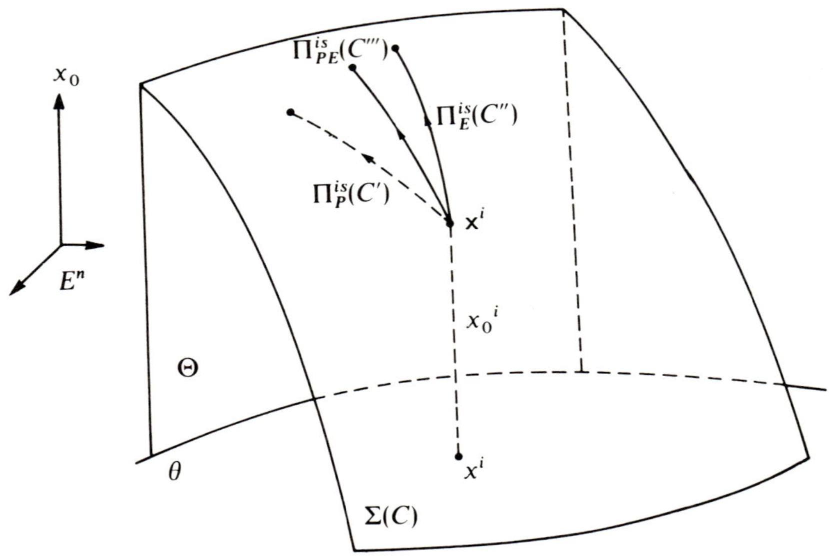

FIG. 2.4. Paths $\Pi_P^{is}(C')$, $\Pi_E^{is}(C'')$, $\Pi_{PE}^{is}(C''')$ and game surface through $\mathbf{x}^i$.

Lemma 2.2. *No point of a path $\Pi_E^{is}(C'')$ which emanates from $\mathbf{x}^i$ is a B-point relative to the game surface through $\mathbf{x}^i$.*

Let $\mathbf{x}^j = (x_0^{\,j}, x^j)$ be a point of a path $\Pi_P^{is}(C')$ generated by (r_P^*, r_E), where x^j is its projection on G. Let $\pi_P^{ij} \subset \pi_P^{is}$ be the path in G emanating from x^i, generated by (r_P^*, r_E) and ending at x^j, which has the same supporting curve as π_P^{is}.

First note that, since the family of game surfaces is defined on $\mathscr{X}^*$, $\mathbf{x}^j$ cannot be an A-point relative to any member of the family if $x^j \notin X^*$. Hence let us suppose that $x^j \in X^*$.

By condition (i) in the definition of an optimal strategy pair, there exists a path π_{PE}^{jf} which emanates from x^j, is generated by (r_P^*, r_E^*) and reaches θ at point x^f.

From Assumption 1.1 it follows directly that $\pi_P^{ij} \cup \pi_{PE}^{jf}$ is a path π_P^{if}. Let (r_P^*, ρ_E) be the strategy pair that generates this path.

Path π_P^{if} reaches θ at point x^f, and associated with it is a value of the cost of transfer

$$V(x^i, \theta; r_P^*, \rho_E) = \mathscr{V}(x^i, x^f; r_P^*, \rho_E, \pi_P^{if}) \tag{2.17}$$

In view of the saddle-point condition we see that

$$V(x^i, \theta;\ r_P{}^*, \rho_E) \leqslant V(x^i, \theta;\ r_P{}^*, r_E{}^*) \triangleq V^*(x^i) \tag{2.18}$$

From the additivity property of cost (2.16) we have

$$\mathscr{V}(x^i, x^f; r_P{}^*, \rho_E, \pi_P^{if}) = \mathscr{V}(x^i, x^j; r_P{}^*, r_E, \pi_P^{ij}) + \mathscr{V}(x^j, x^f; r_P{}^*, r_E{}^*, \pi_{PE}^{jf}) \tag{2.19}$$

where

$$\mathscr{V}(x^j, x^f;\ r_P{}^*, r_E{}^*, \pi_{PE}^{jf}) \triangleq V^*(x^j) \tag{2.20}$$

From (2.17), (2.18), (2.19) and (2.20) we obtain

$$\mathscr{V}(x^i, x^j;\ r_P{}^*, r_E, \pi_P^{ij}) \leqslant V^*(x^i) - \quad {}^*(x^j)$$

Then it follows from (2.13) and from the additivity property of cost that, along path $\Pi_P^{is}(C')$,

$$x_0{}^j - x_0{}^i = \mathscr{V}(x^i, x^j;\ r_P{}^*, r_E, \pi_P^{ij})$$

and hence

$$x_0{}^j - x_0{}^i \leqslant V^*(x^i) - V^*(x^j) \tag{2.21}$$

Finally, it follows from (2.21), (2.9) and (2.10) that x^j is not an A-point relative to the game surface through x^i.

Lemma 2.2 can be established by similar arguments.

Lemmas 2.1 and 2.2 have the following corollaries.

Corollary 2.1. *All points in $\mathscr{X}^*$ of a path $\Pi_{PE}^{is}(C''')$ which emanates from x^i belong to the game surface through x^i.*

This corollary is a direct consequence of Lemmas 2.1 and 2.2 together with the definitions of A- and B-points relative to a game surface.

Corollary 2.2. *A path $\Pi_P^{is}(C')$ whose initial point is a B-point relative to $\Sigma(C)$ has no A-point relative to $\Sigma(C)$ and, of course, no point in $\Sigma(C)$.*

Corollary 2.3. *A path $\Pi_E^{is}(C'')$ whose initial point is an A-point relative to $\Sigma(C)$ has no B-point relative to $\Sigma(C)$ and, of course, no point in $\Sigma(C)$.*

These corollaries are direct consequences of Lemmas 2.1 and 2.2, respectively, together with the translation property of game surfaces.

2.7 UNIQUENESS OF THE VALUE OF THE GAME

Suppose there exist two optimal strategy pairs (r_P^*, r_E^*) and $(\hat{r}_P^*, \hat{r}_E^*)$; that is, both satisfy conditions (i)–(iii) of Definition 2.2. Upon applying these conditions we arrive at once at

Lemma 2.3. *If $x^i \in X^*$, both (r_P^*, r_E^*) and $(\hat{r}_P^*, \hat{r}_E^*)$ are optimal on X^*, and if*

$$(r_P^*, \hat{r}_E^*) \in \mathscr{T}(x^i) \quad \text{and} \quad (\hat{r}_P^*, r_E^*) \in \mathscr{T}(x^i),$$

then

$$V(x^i, \theta;\ r_P^*, r_E^*) = V(x^i, \theta;\ \hat{r}_P^*, \hat{r}_E^*)$$

In other words, the Value of the game $V^*(x^i)$ is independent of the choice of optimal strategy pair.

We also have

Corollary 2.4. *If both (r_P^*, r_E^*) and $(\hat{r}_P^*, \hat{r}_E^*)$ are optimal on X^*, and if $(r_P^*, \hat{r}_E^*)$* and *$(\hat{r}_P^*, r_E^*)$ are playable for all $x^i \in X$, then $(r_P^*, \hat{r}_E^*)$ and $(\hat{r}_P^*, r_E^*)$ are also optimal on X^*.*

III *Differential Quantitative Games*

3.1 THE STATE EQUATIONS AND STRATEGIES

Thus far we have not specified the rules of the game. In this chapter we shall restrict the analysis to differential quantitative games. Let there be given the set of differential equations

$$dx_\nu/dt = f_\nu(x_1, x_2, \ldots, x_n, u_1, u_2, \ldots, u_r, v_1, v_2, \ldots, v_s), \qquad \nu = 1, 2, \ldots, n \tag{3.1}$$

termed *state equations*. Hereafter, $x_n = t$ will denote the time variable. It follows that

$$f_n(x_1, x_2, \ldots, x_n, u_1, u_2, \ldots, u_r, v_1, v_2, \ldots, v_s) \equiv 1$$

We shall assume that state $x = (x_1, x_2, \ldots, x_n)$ belongs to the domain G of E^n, that vectors $u = (u_1, u_2, \ldots, u_r)$ and $v = (v_1, v_2, \ldots, v_s)$ belong to the domains U and V of E^r and E^s, respectively, and that the functions f_ν, $\nu = 1, 2, \ldots, n - 1$, are of class C^1 on $G \times U \times V$.

We shall consider strategies for players J_P and J_E to be functions of x, p: $x \to p(x)$ and e: $x \to e(x)$, $x \in G$, respectively, belonging to prescribed classes of functions, and such that

$$p(x) \in K_u(x) \subseteq U$$

$$e(x) \in K_v(x) \subseteq V$$

where $K_u(x)$ and $K_v(x)$ are given subsets of E^r and E^s, respectively, which may depend on x.

These sets of strategies will be denoted by $\mathscr{S}_P$ and $\mathscr{S}_E$, respectively.

Of course, we can also use the notation of Chapter 1; that is,

$$R_P = \mathscr{S}_P, \qquad R_E = \mathscr{S}_E$$
$$r_P = p, \qquad r_E = e$$

The change in notation is introduced because we wish to emphasize that the sets of strategies are now sets of functions.

Here we need not specify the classes of functions p and e. However, we shall require that $\mathscr{S}_P$ and $\mathscr{S}_E$ satisfy

Assumption 3.1. Whatever $p', p'' \in \mathscr{S}_P$ and $e', e'' \in \mathscr{S}_E$, and whatever $x^j \in G$, functions $p''': \ x \to p'''(x)$ and $e''': \ x \to e'''(x)$, $x \in G$, where

$$\left.\begin{aligned} p'''(x) &= p'(x) \\ e'''(x) &= e'(x) \end{aligned}\right\} \quad \text{for} \quad x_n \leqslant x_n{}^j$$

$$\left.\begin{aligned} p'''(x) &= p''(x) \\ e'''(x) &= e''(x) \end{aligned}\right\} \quad \text{for} \quad x_n > x_n{}^j$$

are strategies; that is, $p''' \in \mathscr{S}_P$ and $e''' \in \mathscr{S}_E$.

3.2 TRANSFER TIME

Definition 3.1. In the following discussion we shall use a variable τ, called *transfer time*, which is a function of time t, $\tau: \ t \to \tau(t)$ on an interval that depends on G, defined by two conditions; namely, $d\tau/dt = 1$ and, for $x^i \in G$, $\tau(t_i) = 0$, where $x_n{}^i = t_i$.

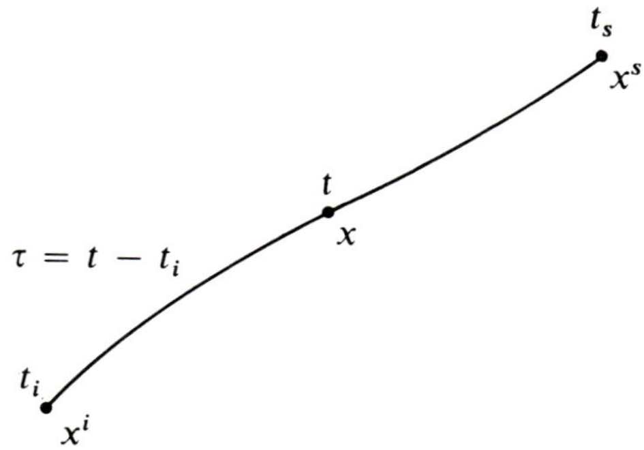

FIG. 3.1. Transfer time τ.

3.3 RULES OF THE GAME

Equation (3.1) can be rewritten as

$$dx/d\tau = f(x, u, v) \tag{3.2}$$

For given strategies p and e, Eq. (3.2) is replaced by

$$dx/d\tau = f(x, p(x), e(x)) \tag{3.3}$$

If $\mathbf{x}$: $\tau \to x = \mathbf{x}(\tau)$, $\tau \in [0, \tau_s)$, is a solution of Eq. (3.3), satisfying $x^i = \mathbf{x}(0)$, then u and v are given by functions of τ, $\mathbf{u}$, and $\mathbf{v}$; namely,

$$\begin{aligned} \mathbf{u}:\quad \tau \to u = \mathbf{u}(\tau), \qquad \mathbf{u}(\tau) = p(\mathbf{x}(\tau)) \\ \mathbf{v}:\quad \tau \to v = \mathbf{v}(\tau), \qquad \mathbf{v}(\tau) = e(\mathbf{x}(\tau)) \\ \tau \in [0, \tau_s) \end{aligned} \tag{3.4}$$

There may be more than one solution of Eq. (3.3) that satisfies $\mathbf{x}(0) = x^i$. Of course, if p and e are C^1 functions of x in a neighborhood of x^i, it follows that there is a unique solution $\mathbf{x}$ of (3.3) satisfying $\mathbf{x}(0) = x^i$ on an interval $\Delta\tau$ containing $\tau = 0$. But if x^i is a point of discontinuity of p or e, or of both p and e, there may be more than one solution of (3.3) that emanates from x^i. Furthermore, a solution that is unique in a neighborhood of the initial time may bifurcate at some later time, if the trajectory reaches a point of discontinuity of p or e, or of both. Also, there may be *no* solution of Eq. (3.3) in $[0, \infty)$ that satisfies $\mathbf{x}(0) = x^i$.

The rules of the game consist of G, $\mathscr{S}_P$, $\mathscr{S}_E$, and mapping γ defined as in Example 1.1 of Section 1.2.

3.4 TARGET θ

Assumption 3.2. Target θ is *closed* and its boundary†

$$\partial\theta \triangleq \bar{\theta} \cap \overline{\text{comp}\ \theta}$$

is defined by the equation

$$\boldsymbol{\theta}(x) = 0 \tag{3.5}$$

where function $\boldsymbol{\theta}$ is of class C^1 and grad $\boldsymbol{\theta}(x) \neq 0$ on a domain containing $\partial\theta$.

† Comp θ means the complement of θ in G. $\bar{\theta}$ and $\overline{\text{comp}\ \theta}$ are the closures of θ and comp θ, respectively, in the topology induced by E^n on G (see Appendix).

3.5 INTEGRAL PERFORMANCE INDEX

Now we shall introduce a functional which assigns a cost to a transfer of the state. Let f_0: $(x, u, v) \rightarrow f_0(x, u, v)$ be a prescribed function of class C^1 on $G \times U \times V$.

In addition to Assumption 3.1, we now introduce

Assumption 3.3. The sets of strategies $\mathscr{S}_P$ and $\mathscr{S}_E$ are such that, for all $x^i \in G$, for all $p \in \mathscr{S}_P$, for all $e \in \mathscr{S}_E$, and for all paths π^{is} in G emanating from x^i, generated by (p, e), represented by $\mathbf{x}$: $\tau \rightarrow x = \mathbf{x}(\tau)$, $\tau \in [0, \tau_s]$, the integral

$$\int_0^{\tau_s} f_0(\mathbf{x}(\tau), \mathbf{u}(\tau), \mathbf{v}(\tau))\, d\tau$$

where $\mathbf{u}(\tau) = p(\mathbf{x}(\tau))$, $\mathbf{v}(\tau) = e(\mathbf{x}(\tau))$, is defined.

Now let us consider a functional or performance index in integral form

$$\mathscr{V}(x^i, x^s; p, e, \pi^{is}) \triangleq \int_0^{\tau_s} f_0(\mathbf{x}(\tau), \mathbf{u}(\tau), \mathbf{v}(\tau))\, d\tau \tag{3.6}$$

where $x^s = \mathbf{x}(\tau_s)$.

If $\tau_s = 0$, then π^{is} is a null path π^0, and

$$\mathscr{V}(x^i, x^i;\ p, e, \pi^0) = 0 \qquad \forall p \in \mathscr{S}_P \quad \forall e \in \mathscr{S}_E \tag{3.7}$$

Hence, Assumption 2.1 is satisfied.

If $x^s \in \theta$, according to our earlier notation, we have $\tau_s = \tau_f$ and

$$V(x^i, \theta; p, e) \triangleq \int_0^{\tau_f} f_0(\mathbf{x}(\tau), \mathbf{u}(\tau), \mathbf{v}(\tau))\, d\tau \tag{3.8}$$

Let us introduce a variable x_0 and a function $\mathbf{x}_0$: $\tau \rightarrow x_0 = \mathbf{x}_0(\tau)$, $\tau \in [0, \tau_s]$, such that

$$\mathbf{x}_0(\tau) + \int_\tau^{\tau_s} f_0(\mathbf{x}(\xi), \mathbf{u}(\xi), \mathbf{v}(\xi))\, d\xi = C \tag{3.9}$$

where C is any given constant.

If $\tau_s \neq 0$ we have

$$dx_0/d\tau = f_0(x, u, v) \tag{3.10}$$

for all $\tau \in (0, \tau_s)$.

Equations (3.2) and (3.10) constitute a set of $n + 1$ scalar equations which we shall write in vector form

$$d\mathsf{x}/d\tau = \mathsf{f}(\mathsf{x}, u, v) \tag{3.11}$$

where $\mathsf{x} = (x_0, x)$ is the state in $\mathscr{G} \triangleq G \times \{x_0\}$ and

$$\mathsf{f}(\mathsf{x}, u, v) \triangleq (f_0(x, u, v), f_1(x, u, v), \ldots, f_n(x, u, v))$$

3.6 A VARIATIONAL EQUATION AND ITS ADJOINT

Next we introduce

Assumption 3.4. There exists a strategy pair (p^*, e^*) that is optimal on a set $X \subseteq G$.

By (3.7), (p^*, e^*) is optimal on $X^* \triangleq X \cup \theta$. Consider a non-null optimal path $\Pi^*(C)$ that emanates from $\mathsf{x}^i \in \mathscr{X}^*$, $\mathscr{X}^* \triangleq X^* \times \{x_0\}$, is

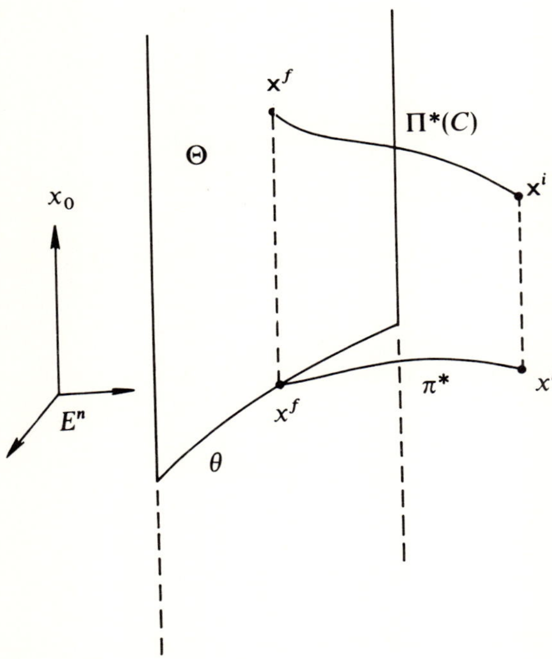

FIG. 3.2. Path $\Pi^*(C)$ and its projection π^* on G.

generated by (p^*, e^*), is given by $\mathbf{x}^*$: $\tau \to x = \mathbf{x}^*(\tau)$ on $[0, \tau_f]$, and reaches $\Theta \triangleq \theta \times \{x_0\}$ at $x^f = \mathbf{x}^*(\tau_f)$. In other words, $\mathbf{x}^*$ is a solution of

$$dx/d\tau = f(x, p^*(x), e^*(x)) \tag{3.12}$$

Let π^* be the projection of $\Pi^*(C)$ on G.

Definition 3.2. If there exists a neighborhood of π^* on which p^* and e^*, are of class C^1, we shall associate with Eq. (3.12) the *variational equation*

$$\frac{d\eta}{d\tau} = \left(\frac{\partial f}{\partial x} + \frac{\partial f}{\partial u}\frac{\partial p^*}{\partial x} + \frac{\partial f}{\partial v}\frac{\partial e^*}{\partial x}\right)\Bigg|_{x=\mathbf{x}^*(\tau)} \eta \tag{3.13}$$

where

$$\frac{\partial f}{\partial x} \triangleq \left[\frac{\partial f_\nu(x, u, v)}{\partial x_\alpha}\right] \qquad \begin{aligned}\nu &= 0, 1, \ldots, n\\ \alpha &= 0, 1, \ldots, n\end{aligned}$$

$$\frac{\partial f}{\partial u} \triangleq \left[\frac{\partial f_\nu(x, u, v)}{\partial u_\alpha}\right] \qquad \begin{aligned}\nu &= 0, 1, \ldots, n\\ \alpha &= 1, 2, \ldots, r\end{aligned}$$

$$\frac{\partial f}{\partial v} \triangleq \left[\frac{\partial f_\nu(x, u, v)}{\partial v_\alpha}\right] \qquad \begin{aligned}\nu &= 0, 1, \ldots, n\\ \alpha &= 1, 2, \ldots, s\end{aligned}$$

evaluated for $u = p^*(x)$, $v = e^*(x)$; and

$$\frac{\partial p^*}{\partial x} \triangleq \left[\frac{\partial p_\nu^*(x)}{\partial x_\alpha}\right] \qquad \begin{aligned}\nu &= 1, 2, \ldots, r\\ \alpha &= 0, 1, \ldots, n\end{aligned}$$

$$\frac{\partial e^*}{\partial x} \triangleq \left[\frac{\partial e_\nu^*(x)}{\partial x_\alpha}\right] \qquad \begin{aligned}\nu &= 1, 2, \ldots, s\\ \alpha &= 0, 1, \cdots, n\end{aligned}$$

$$\eta \triangleq \begin{pmatrix}\eta_0\\ \eta_1\\ \cdot\\ \cdot\\ \cdot\\ \eta_n\end{pmatrix}$$

For given initial condition $\eta = \eta^i$ at $\tau = 0$, the solution $\boldsymbol{\eta}$: $\tau \to \eta = \boldsymbol{\eta}(\tau)$ of Eq. (3.13) is unique and continuous on $[0, \tau_f]$. Furthermore, $\boldsymbol{\eta}(\tau)$, $\tau \in [0, \tau_f]$, is nonzero provided η^i is nonzero.

Definition 3.3. The equation adjoint to Eq. (3.13) is

$$\frac{d\lambda}{d\tau} = -\left(\frac{\partial f}{\partial x} + \frac{\partial f}{\partial u}\frac{\partial p^*}{\partial x} + \frac{\partial f}{\partial v}\frac{\partial e^*}{\partial x}\right)^{\mathrm{T}}\Bigg|_{\mathbf{x}=\mathbf{x}^*(\tau)} \lambda \tag{3.14}$$

where

$$\lambda \triangleq \begin{pmatrix} \lambda_0 \\ \lambda_1 \\ \cdot \\ \cdot \\ \cdot \\ \lambda_n \end{pmatrix}$$

For given initial condition $\lambda = \lambda^i$ at $\tau = 0$, the solution $\boldsymbol{\lambda}$: $\tau \to \lambda = \boldsymbol{\lambda}(\tau)$ of Eq. (3.14) is unique and continuous on $[0, \tau_f]$. Furthermore, $\boldsymbol{\lambda}(\tau)$ is nonzero provided λ^i is nonzero.

Equations (3.13) and (3.14) can be developed as follows:

$$\frac{d\eta_\nu}{d\tau} = \sum_{\alpha=0}^{n}\left(\frac{\partial f_\nu}{\partial x_\alpha} + \sum_{k=1}^{r}\frac{\partial f_\nu}{\partial u_k}\frac{\partial p_k^*}{\partial x_\alpha} + \sum_{l=1}^{s}\frac{\partial f_\nu}{\partial v_l}\frac{\partial e_l^*}{\partial x_\alpha}\right)\Bigg|_{\mathbf{x}=\mathbf{x}^*(\tau)} \eta_\alpha \tag{3.15}$$

and

$$\frac{d\lambda_\nu}{d\tau} = -\sum_{\alpha=0}^{n}\left(\frac{\partial f_\alpha}{\partial x_\nu} + \sum_{k=1}^{r}\frac{\partial f_\alpha}{\partial u_k}\frac{\partial p_k^*}{\partial x_\nu} + \sum_{l=1}^{s}\frac{\partial f_\alpha}{\partial v_l}\frac{\partial e_l^*}{\partial x_\nu}\right)\Bigg|_{\mathbf{x}=\mathbf{x}^*(\tau)} \lambda_\alpha \tag{3.16}$$

$$\nu = 0, 1, 2, \ldots, n$$

It follows from (3.13) and (3.14) that

$$(d/d\tau)(\boldsymbol{\lambda}(\tau)\cdot\boldsymbol{\eta}(\tau)) = 0$$

so that

$$\boldsymbol{\lambda}(\tau)\cdot\boldsymbol{\eta}(\tau) = \text{constant} \qquad \forall\tau \in [0, \tau_f] \tag{3.17}$$

If the initial vectors $\lambda^i = \boldsymbol{\lambda}(0)$ and $\eta^i = \boldsymbol{\eta}(0)$ are such that

$$\lambda^i \cdot \eta^i = 0 \qquad \text{with} \qquad \|\lambda^i\|\ \|\eta^i\| \neq 0$$

that is, if λ^i is perpendicular to η^i, then it follows from (3.17) that

$$\boldsymbol{\lambda}(\tau)\cdot\boldsymbol{\eta}(\tau) = 0 \qquad \forall\tau \in [0, \tau_f] \tag{3.18}$$

and since $\|\boldsymbol{\lambda}(\tau)\|\ \|\boldsymbol{\eta}(\tau)\| \neq 0$, $\boldsymbol{\lambda}(\tau)$ is perpendicular to $\boldsymbol{\eta}(\tau)$ for all $\tau \in [0, \tau_f]$.

Definition 3.4. Equation (3.13) defines a linear transformation $\mathscr{A}(0, \tau)$ of η^i such that

$$\boldsymbol{\eta}(\tau) = \mathscr{A}(0, \tau)\eta^i \tag{3.19}$$

Since Eq. (3.13) is linear and homogeneous, this transformation is nonsingular; that is, the inverse transformation $\mathscr{A}^{-1}(0, \tau)$ such that

$$\eta^i = \mathscr{A}^{-1}(0, \tau)\boldsymbol{\eta}(\tau)$$

is defined for all $\tau \in [0, \tau_f]$.

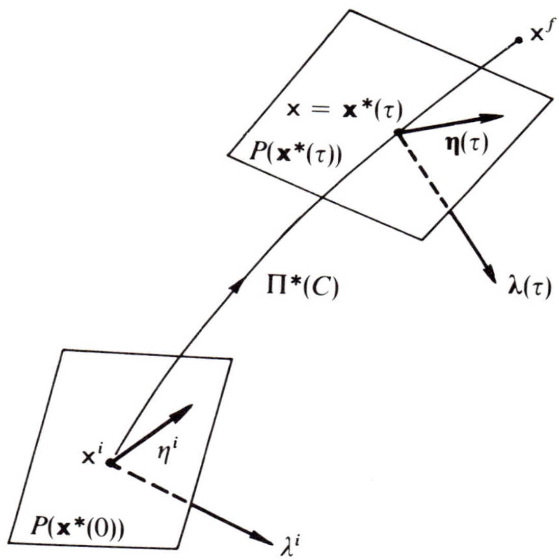

FIG. 3.3. Transform of plane $P(\mathbf{x}^*(0))$ due to $\mathscr{A}(0, \tau)$.

Next, let $P(\mathbf{x}^*(0))$ be an n-dimensional plane containing point $\mathbf{x}^*(0)$ of $\Pi^*(C)$, and let $P(\mathbf{x}^*(\tau))$ be its transform due to the linear transformation $\mathscr{A}(0, \tau)$; namely,

$$P(\mathbf{x}^*(\tau)) = \mathscr{A}(0, \tau)P(\mathbf{x}^*(0)) \triangleq \{\mathbf{x}^*(\tau) + \mathscr{A}(0, \tau)\eta^i : \quad \mathbf{x}^*(0) + \eta^i \in P(\mathbf{x}^*(0))\}$$

It follows from (3.18) that, if λ^i is perpendicular to $P(\mathbf{x}^*(0))$, then $\boldsymbol{\lambda}(\tau)$ is perpendicular to $P(\mathbf{x}^*(\tau))$ for all $\tau \in [0, \tau_f]$.

3.7 JOINING OF PATHS

We shall now prove that Assumption 1.1 is satisfied.

Consider a trajectory Γ^{ij} in G that emanates from x^i, is generated by (p, e), is given by $\mathbf{x}$: $\tau \to x = \mathbf{x}(\tau)$ on $[0, \tau_j]$, and ends at point

$$x^j = \mathbf{x}(\tau_j).$$

Consider also describing curve $\Gamma_{\hat{\mathbf{x}}}(x^j, \hat{p}, \hat{e})$ that emanates from x^j and is given by $\hat{\mathbf{x}} \in \gamma(x^j, \hat{p}, \hat{e})$. Of course, $\mathbf{x}(\tau_j) = \hat{\mathbf{x}}(0) = x^j$.

If Γ^{ij}, or $\Gamma_{\hat{\mathbf{x}}}(x^j, \hat{p}, \hat{e})$, or both, reduces to a single point, Assumption 1.1 is trivially satisfied. Accordingly, let us suppose that neither Γ^{ij} nor $\Gamma_{\hat{\mathbf{x}}}(x^j, \hat{p}, \hat{e})$ reduces to a single point. Then, $\mathbf{x}$ is a solution of

$$dx/d\tau = f(x, p(x), e(x)) \tag{3.20}$$

with initial condition $\mathbf{x}(0) = x^i$; and $\hat{\mathbf{x}}$ is a maximal solution of

$$dx/d\tau = f(x, \hat{p}(x), \hat{e}(x)) \tag{3.21}$$

with initial condition $\hat{\mathbf{x}}(0) = x^j$.

Now consider the (n-1)-dimensional plane $P(x^j)$ in E^n which contains point x^j and is perpendicular to the time axis, namely the plane whose equation is $x_n = x_n^{\ j}$. This plane separates E^n into two half-spaces, the closed half-space $D_1 \triangleq \{x\colon\ x_n \leqslant x_n^{\ j}\}$ and the open half-space $D_2 \triangleq \{x\colon\ x_n > x_n^{\ j}\}$.

From Assumption 3.1 it follows that functions $\tilde{p}$: $x \to \tilde{p}(x)$ and $\tilde{e}$: $x \to \tilde{e}(x)$, $x \in G$, where

$$\left.\begin{aligned}\tilde{p}(x) &= p(x)\\ \tilde{e}(x) &= e(x)\end{aligned}\right\}\qquad \text{for}\qquad x \in D_1 \cap G$$

$$\left.\begin{aligned}\tilde{p}(x) &= \hat{p}(x)\\ \tilde{e}(x) &= \hat{e}(x)\end{aligned}\right\}\qquad \text{for}\qquad x \in D_2 \cap G$$

are strategies.

Moreover, if the players employ the strategy pair $(\tilde{p}, \tilde{e})$, the state equation becomes

$$dx/d\tau = f(x, \tilde{p}(x), \tilde{e}(x)) \tag{3.22}$$

There exists a maximal solution $\tilde{\mathbf{x}}$: $\tau \to x = \tilde{\mathbf{x}}(\tau)$ of Eq. (3.22) which satisfies the initial condition $\tilde{\mathbf{x}}(0) = x^i$, namely the solution $\tilde{\mathbf{x}}$ such that

$$\tilde{\mathbf{x}}(\tau) = \mathbf{x}(\tau) \qquad \text{for} \qquad \tau \in [0, \tau_j]$$

$$\tilde{\mathbf{x}}(\tau) = \hat{\mathbf{x}}(\tau - \tau_j) \qquad \text{for} \qquad \tau \in [\tau_j, \tau_j + \tau^+]$$

where τ^+ is the extreme future value of τ associated with solution $\hat{\mathbf{x}}$. Solution $\tilde{\mathbf{x}}$ generates describing curve $\Gamma_{\tilde{\mathbf{x}}}(x^i, \tilde{p}, \tilde{e})$, and we have

$$\Gamma_{\tilde{\mathbf{x}}}(x^i, \tilde{p}, \tilde{e}) = \Gamma^{ij} \cup \Gamma_{\hat{\mathbf{x}}}(x^j, \hat{p}, \hat{e})$$

Furthermore,

$$\hat{p} = p \Rightarrow \tilde{p} = p$$

and

$$\hat{e} = e \Rightarrow \tilde{e} = e$$

Since these arguments apply whatever the trajectory Γ^{ij} and the describing curve $\Gamma_{\hat{\mathbf{x}}}(x^j, \hat{p}, \hat{e})$, Assumption 1.1 is satisfied. One can readily verify that Assumption 2.2 is also satisfied.

Hence, Lemmas 2.1 and 2.2 apply to differential games for which Assumption 3.1 is satisfied.

3.8 TWO FAMILIES OF PATHS

Again consider a non-null optimal path $\Pi^*(C)$ that emanates from $\mathbf{x}^i \in \mathscr{X}^*$, is generated by (p^*, e^*), and reaches Θ at $\mathbf{x}^f$. Let π^* be its projection on G.

Next we introduce

Assumption 3.5. All points of π^*, with the possible exception of its terminal point, are interior points of X^*.

Assumption 3.6. For all $x \in \pi^*$, and for all $u \in K_u(x)$ and $v \in K_v(x)$, there exist strategies $\alpha_x \in \mathscr{S}_P$ and $\beta_x \in \mathscr{S}_E$ which are of class C^1 on some neighborhood of x in G, and which satisfy the conditions $\alpha_x(x) = u$ and $\beta_x(x) = v$, respectively.

Let $\mathbf{x}^j \triangleq (x_0{}^j, x^j)$ be any point of $\Pi^*(C)$, different from terminal point $\mathbf{x}^f$. From Assumption 3.5 it follows that $\mathbf{x}^j$ is an interior point of $\mathscr{X}^*$.

Let us also consider paths $\Pi_E^{jl}(C')$ and $\Pi_P^{jm}(C'')$, which emanate from

x^j and belong to $\mathscr{X}^*$; namely,

$$\Pi_E^{jl}(C') \triangleq \{x^k: \ x_0{}^k + \mathscr{V}(x^k, x^l; \alpha_{x^j}, e^*, \pi_E^{kl}) = C', \quad \pi_E^{kl} \subset \pi_E^{jl}\} \tag{3.23}$$

$$\Pi_P^{jm}(C'') \triangleq \{x^k: \ x_0{}^k + \mathscr{V}(x^k, x^m; p^*, \beta_{x^j}, \pi_P^{km}) = C'', \quad \pi_P^{km} \subset \pi_P^{jm}\} \tag{3.24}$$

where α_{x^j} and β_{x^j} are strategies satisfying Assumption 3.6 at point $x^j \in \pi^*$.

Let

$$\alpha_{x^j}(x^j) = u^b \in K_u(x^j)$$
$$\beta_{x^j}(x^j) = v^b \in K_v(x^j)$$

where u^b and v^b are given vectors.

According to Corollary 2.1, $\Pi^*(C)$ belongs to the game surface through x^i; hence it follows from the definitions of a game surface and of an optimal path that the game surface through x^i is $\Sigma(C)$.

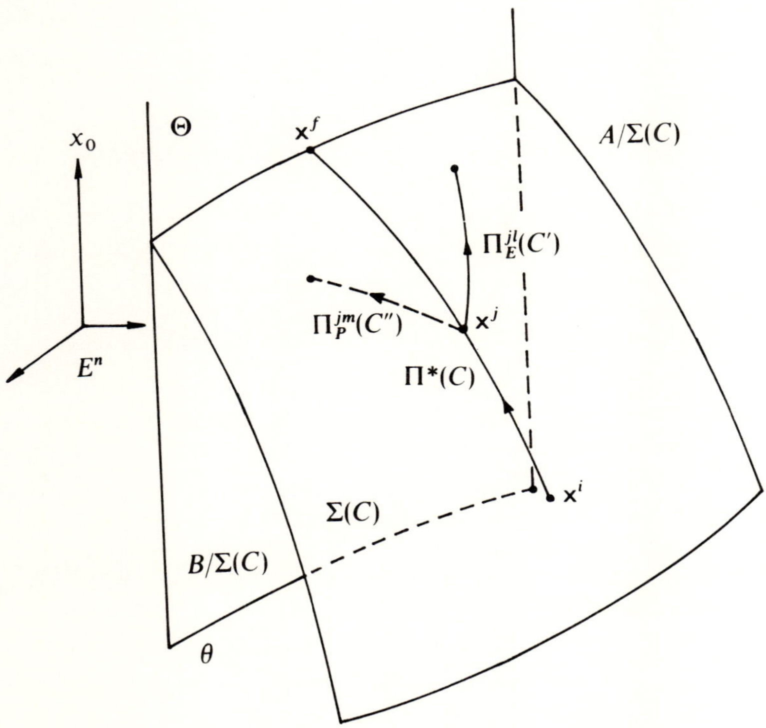

FIG. 3.4. Paths $\Pi_E^{jl}(C')$ and $\Pi_P^{jm}(C'')$.

Consequently,

$$x^j \in \Sigma(C)$$

Now it follows from Lemmas 2.1 and 2.2 that

(i) no point of $\Pi_E^{jl}(C')$ is a B-point relative to $\Sigma(C)$; and

(ii) no point of $\Pi_P^{jm}(C'')$ is an A-point relative to $\Sigma(C)$.

Finally, since $\Pi_E^{jl}(C')$ and $\Pi_P^{jm}(C'')$ belong to $\mathscr{X}^*$, we have

$$\Pi_E^{jl}(C') \subset \widetilde{A/\Sigma}(C) \quad \text{where} \quad \widetilde{A/\Sigma}(C) \triangleq (A/\Sigma(C)) \cup \Sigma(C) \tag{3.25}$$

$$\Pi_P^{jm}(C'') \subset \widetilde{B/\Sigma}(C) \quad \text{where} \quad \widetilde{B/\Sigma}(C) \triangleq (B/\Sigma(C)) \cup \Sigma(C) \tag{3.26}$$

Next we shall consider two families of paths emanating from points $x^j \in \Pi^*(C)$, $x^j \notin \Theta$; namely,

1. the set $\{\Pi_E^{jl}(C')\}$

and

2. the set $\{\Pi_P^{jm}(C'')\}$,

generated by strategies α_{x^j} and β_{x^j}, respectively, for all $u^b \in K_u(x^j)$ and all $v^b \in K_v(x^j)$.

3.9 REGULAR INTERIOR POINTS OF A GAME SURFACE

Before discussing other properties of paths $\Pi_E^{jl}(C')$ and $\Pi_P^{jm}(C'')$, let us return to the definition of game surface $\Sigma(C)$ and recall its equation

$$\Phi(x) \triangleq x_0 + V^*(x) = C, \qquad x \in X^* \tag{3.27}$$

Then it follows from the definition of $A/\Sigma(C)$ and $B/\Sigma(C)$ that

(i) at every point x of $A/\Sigma(C)$,

$$\Phi(x) > C$$

and

(ii) at every point x of $B/\Sigma(C)$,

$$\Phi(x) < C$$

Let $x^j + \delta^E x$ and $x^j + \delta^P x$ denote points of paths $\Pi_E^{jl}(C')$ and $\Pi_P^{jm}(C'')$, respectively; and let ξ and ζ be functions

$$\xi\colon \quad \tau \to \delta^E x = \xi(\tau)$$

$$\zeta\colon \quad \tau \to \delta^P x = \zeta(\tau)$$

where $\xi(0) = \zeta(0) = 0$ at point x^j, and the intervals of definition depend on α_{x^j} and β_{x^j}, respectively.

From (3.25) and (3.26), and from (i) and (ii) above, we can conclude that

1. at every point $x^j + \delta^E x$ of path $\Pi_E^{jl}(C')$,

$$\Phi(x^j + \delta^E x) \geqslant C \tag{3.28}$$

and

2. at every point $x^j + \delta^P x$ of path $\Pi_P^{jm}(C'')$,

$$\Phi(x^j + \delta^P x) \leqslant C \tag{3.29}$$

Next we introduce

Assumption 3.7. At every point x^j of π^* there exists a neighborhood $\Delta(x^j)$ in G of x^j, on which p^* and e^* are of class C^1.

Assumption 3.8.† At every point x^j of $\Pi^*(C)$, $x^j \notin \Theta$, there exists a neighborhood $\Delta(x^j)$ in $\mathscr{X}^*$ of x^j, such that the function Φ: $x \rightarrow \Phi(x)$, $x \in \Delta(x^j)$, is of class C^1.

In view of Assumptions 3.6 and 3.7 we have

PROPERTY 3.1. Whatever $x^j \in \Pi^*(C)$, $x^j \notin \Theta$, there exist *non-null* paths $\Pi_E^{jl}(C')$ and $\Pi_P^{jm}(C'')$, and an interval $[0, \epsilon]$, $\epsilon > 0$, on which the functions ξ and ζ are continuous.

Then, as a consequence of Assumption 3.8 we have

PROPERTY 3.2. Every point x of game surface $\Sigma(C)$, in a neighborhood in $\mathscr{X}^*$ of x^j, can be represented by

$$x = x^j + \epsilon\eta^j + o(\epsilon)$$

where $\lim_{\epsilon\to 0} [\|o(\epsilon)\|/\epsilon] = 0$ and η^j belongs to an n-dimensional plane, the tangent plane of the surface at x^j.

Definition 3.5. If a point x^j of $\Sigma(C)$ possesses a neighborhood that belongs to $\mathscr{X}^*$, it will be called an *interior point* of $\Sigma(C)$. A point x^j of $\Sigma(C)$ at which Property 3.2 holds will be called a *regular point* of $\Sigma(C)$. An optimal path $\Pi^*(C)$ all of whose points, with the possible exception of the terminal one, are regular interior points of $\Sigma(C)$ will be called a *regular optimal path*.

† One can verify that Assumption 3.8 is a consequence of Assumptions 3.5 and 3.7 if π^* is not tangent to θ.

Note that Assumption 3.8 is stronger than the assumption of regularity; of course, according to Assumption 3.8 together with (3.27), the partial derivatives

$$\frac{\partial V^*(x)}{\partial x_1} \cdots \frac{\partial V^*(x)}{\partial x_n}$$

are defined and continuous, and hence

$$\operatorname{grad} \Phi(\mathbf{x}) = \left(1, \frac{\partial V^*(x)}{\partial x_1}, \ldots, \frac{\partial V^*(x)}{\partial x_n}\right) \tag{3.30}$$

is defined and continuous at every point $\mathbf{x}^j$, $\mathbf{x}^j \notin \Theta$, of $\Pi^*(C)$. However, at a regular point of $\Sigma(C)$, some of the partial derivatives of $V^*(x)$ may be not defined, since the tangent plane of $\Sigma(C)$ at that point may be parallel to the x_0-axis.

Now let us return to path $\Pi_E^{jl}(C')$, defined in Section 3.8, that emanates from $\mathbf{x}^j \in \Pi^*(C)$, $\mathbf{x}^j \notin \Theta$, and is generated by strategy pair (α_{x^j}, e^*). We shall assume that $\Pi_E^{jl}(C')$ is *not* a null path.†

It follows from Property 3.1 that

$$\|\delta^E \mathbf{x}\| \to 0 \qquad \text{as} \qquad \tau \to 0$$

Since $\mathbf{x}^j \in \Sigma(C)$, and accordingly $\Phi(\mathbf{x}^j) = C$, it follows from (3.28) and from Assumption 3.8 that

$$\operatorname{grad} \Phi(\mathbf{x}^j) \cdot \delta^E \mathbf{x} + o(\|\delta^E \mathbf{x}\|) \geqslant 0 \tag{3.31}$$

where

$$\lim_{\|\delta^E \mathbf{x}\| \to 0} o \frac{(\|\delta^E \mathbf{x}\|)}{\|\delta^E \mathbf{x}\|} = 0$$

Upon dividing (3.31) by τ, which is positive for $\delta^E \mathbf{x} \neq 0$, we get

$$\operatorname{grad} \Phi(\mathbf{x}^j) \cdot \frac{\delta^E \mathbf{x}}{\tau} + \frac{o(\|\delta^E \mathbf{x}\|)}{\tau} \geqslant 0$$

Then allowing $\tau \to 0$, and consequently

$$(\delta^E \mathbf{x}/\tau) \to \mathbf{f}(\mathbf{x}^j, u^b, e^*(x^j))$$

we obtain

$$\operatorname{grad} \Phi(\mathbf{x}^j) \cdot \mathbf{f}(\mathbf{x}^j, u^b, e^*(x^j)) \geqslant 0 \tag{3.32}$$

for all $u^b \in K_u(x^j)$, and for all $\mathbf{x}^j \in \Pi^*(C)$ with the possible exception of the terminal point of $\Pi^*(C)$.

† Such non-null paths exist according to Assumptions 3.6 and 3.7.

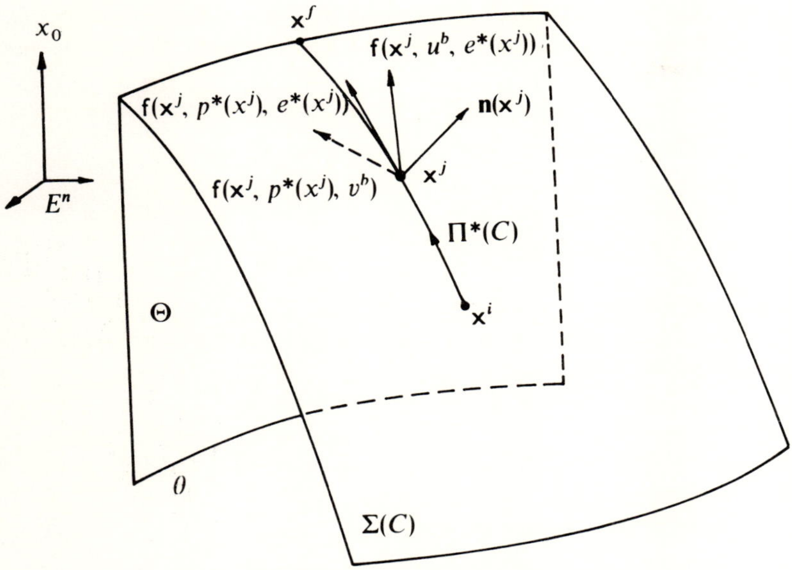

FIG. 3.5. Regular interior points of a game surface.

By invoking similar arguments for a non-null path $\Pi_P^{jm}(C'')$ that emanates from point $\mathbf{x}^j \in \Pi^*(C)$, $\mathbf{x}^j \notin \Theta$, and is generated by strategy pair (p^*, β_{x^j}), we obtain

$$\text{grad } \Phi(\mathbf{x}^j) \cdot \mathbf{f}(\mathbf{x}^j, p^*(x^j), v^b) \leqslant 0 \tag{3.33}$$

for all $v^b \in K_v(x^j)$, and for all $\mathbf{x}^j \in \Pi^*(C)$ with the possible exception of the terminal point of $\Pi^*(C)$.

Moreover, since $\Pi^*(C) \subset \Sigma(C)$, we have

$$\text{grad } \Phi(\mathbf{x}^j) \cdot \mathbf{f}(\mathbf{x}^j, p^*(x^j), e^*(x^j)) = 0 \tag{3.34}$$

for all $\mathbf{x}^j$ which belong to $\Pi^*(C)$, with the possible exception of the terminal point of $\Pi^*(C)$.

Let

$$\mathbf{n}(\mathbf{x}^j) \triangleq \text{grad } \Phi(\mathbf{x}^j), \qquad \mathbf{x}^j \notin \Theta$$

It follows from (3.30) that

$$\mathbf{n}_0(\mathbf{x}^j) = 1 \tag{3.35}$$

where $\mathbf{n}_0(\mathbf{x}^j)$ is the x_0-component of $\mathbf{n}(\mathbf{x}^j)$. Note that $\mathbf{n}(\mathbf{x}^j)$ is normal to the game surface $\Sigma(C)$ and directed into region $A/\Sigma(C)$ at point x^j of $\Sigma(C)$; that is, there exists a positive number σ such that point $\mathrm{x}^j + \epsilon\mathbf{n}(\mathrm{x}^j)$ belongs to $A/\Sigma(C)$ for all ϵ on $(0, \sigma)$.

3.10 TRANSFORMATION OF A TANGENT PLANE

Again consider regular optimal path $\Pi^*(C)$, represented by $\mathbf{x}^*$:

$$\tau \to \mathbf{x}^*(\tau), \qquad \tau \in [0, \tau_f].$$

Let $\mathrm{x}^i \triangleq \mathbf{x}^*(0)$ and $\mathrm{x}^j \triangleq \mathbf{x}^*(\tau_j)$, $\tau_j \in [0, \tau_f)$.

Now let us consider a neighborhood in $\Sigma(C)$ of point x^i, denoted by $\Delta(\mathrm{x}^i)$, such that

$$\Delta(\mathrm{x}^i) \triangleq \{\mathrm{x}\colon \ \mathrm{x} = \mathbf{x}(0) \in \Sigma(C)\}$$

where

$$\mathbf{x}(0) \triangleq \mathbf{x}^*(0) + \epsilon\eta^i + o(\epsilon) \tag{3.36}$$

and

1. $\lim\limits_{\epsilon \to 0} \left[\dfrac{\|o(\epsilon)\|}{\epsilon}\right] = 0$

2. $\eta^i \in P_\Sigma(\mathbf{x}^*(0))$, the tangent plane of $\Delta(\mathrm{x}^i)$, and hence of $\Sigma(C)$, at $\mathbf{x}^*(0)$.

We shall be interested in a transformation of $\Delta(\mathrm{x}^i)$ by means of state equation (3.12), that is in

$$\Delta(\mathrm{x}^j) \triangleq \{\mathrm{x}\colon \ \mathrm{x} = \mathbf{x}(\tau_j), \qquad \tau_j \in [0, \tau_f)\}$$

where $\mathbf{x}\colon \ \tau \to \mathbf{x}(\tau)$, $\tau \in [0, \tau_j]$ represents a trajectory Γ^{ij} whose supporting curve is a solution of (3.12) with initial condition given by (3.36).

It follows from Assumption 3.7 that, to each initial point $\mathbf{x}(0) \in \Delta(\mathrm{x}^i)$, there corresponds a unique trajectory Γ^{ij} provided that $|\epsilon|$ is sufficiently small. Moreover, since $\tau_j \in [0, \tau_f)$, $\mathbf{x}^*(\tau)$ does not belong to Θ for all $\tau \in [0, \tau_j]$. Since θ is closed, $\mathbf{x}(\tau)$ does not belong to Θ for all $\tau \in [0, \tau_j]$, which implies that trajectory Γ^{ij} is a path, no matter what the initial point $\mathbf{x}(0) \in \Delta(\mathrm{x}^i)$, provided that $|\epsilon|$ is sufficiently small.

From the dependence on initial conditions of the solution $\mathbf{x}\colon \ \tau \to \mathbf{x}(\tau)$, $\tau \in [0, \tau_j]$, it follows that

$$\mathbf{x}(\tau_j) = \mathbf{x}^*(\tau_j) + \epsilon\eta^j + o(\tau_j, \epsilon)$$

where $\|o(\tau_j, \epsilon)\|/\epsilon$ tends to zero uniformly for all τ_j as $\epsilon \to 0$, and $\eta^j \triangleq \boldsymbol{\eta}(\tau_j)$ where $\boldsymbol{\eta}\colon \ \tau \to \eta = \boldsymbol{\eta}(\tau)$ is the solution of variational equation

(3.13) on $[0, \tau_f]$, with initial condition $\boldsymbol{\eta}(0) = \eta^i$; that is,

$$\boldsymbol{\eta}(\tau_j) = \mathscr{A}(0, \tau_j)\eta^i$$

It follows from Corollary 2.1 that

$$\mathbf{x}(\tau_j) = \mathbf{x}^*(\tau_j) + \epsilon\eta^j + o(\tau_j, \epsilon) \in \Sigma(C) \tag{3.37}$$

where $\|o(\tau_j, \epsilon)\|/\epsilon$ tends to zero uniformly for all τ_j as $\epsilon \to 0$.

Since (3.37) is satisfied for all $\mathbf{x}(0) \in \Delta(\mathrm{x}^i)$,

$$\Delta(\mathrm{x}^j) \subset \Sigma(C)$$

Since $\eta^i \in P_\Sigma(\mathbf{x}^*(0))$, and since $\mathscr{A}(0, \tau_j)$ is a linear nonsingular transformation, it follows that $\boldsymbol{\eta}(\tau_j)$ belongs to an n-dimensional plane which contains point $\mathbf{x}^*(\tau_j)$. In view of (3.37) this plane is the tangent plane of $\Sigma(C)$, $P_\Sigma(\mathbf{x}^*(\tau_j))$, at $\mathbf{x}^*(\tau_j)$. Furthermore, the orientation of $P_\Sigma(\mathbf{x}^*(\tau_j))$ varies continuously with τ_j.

3.11 REGULAR OPTIMAL PATHS

Now consider the solution $\boldsymbol{\lambda}$: $\tau \to \lambda = \boldsymbol{\lambda}(\tau)$, $\tau \in [0, \tau_f]$, of adjoint equation (3.14) subject to the initial condition $\boldsymbol{\lambda}(0) = \lambda^i$.

If we choose λ^i perpendicular to the tangent plane of $\Sigma(C)$ at point $\mathbf{x}^*(0)$, that is, if λ^i has the same supporting line as $\mathbf{n}(\mathbf{x}^*(0))$, then, as a consequence of (3.18) together with the conclusion of Section 3.10, $\boldsymbol{\lambda}(\tau_j)$ is perpendicular to the tangent plane of $\Sigma(C)$ at point $\mathbf{x}^*(\tau_j)$; namely, its supporting line is that of $\mathbf{n}(\mathbf{x}^*(\tau_j))$.

It follows from Assumption 3.8, together with the definition of Φ and (3.35), that $\boldsymbol{\rho}$: $\tau_j \to \boldsymbol{\rho}(\tau_j)$, where $\boldsymbol{\rho}(\tau_j) = \mathbf{n}(\mathbf{x}^*(\tau_j))$, is a nonzero continuous vector function of τ_j on $[0, \tau_f)$. Furthermore, $\boldsymbol{\lambda}$: $\tau_j \to \boldsymbol{\lambda}(\tau_j)$ is a nonzero continuous vector function of τ_j on $[0, \tau_f]$. Hence, if we choose λ^i in the same direction as $\mathbf{n}(\mathbf{x}^*(0))$, $\boldsymbol{\lambda}(\tau_j)$ is codirectional with $\mathbf{n}(\mathbf{x}^*(\tau_j))$ for all $\tau_j \in [0, \tau_f)$.

These properties of $\boldsymbol{\lambda}(\tau_j)$, together with (3.32), (3.33) and (3.34), result in

$$\left.\begin{aligned} &\boldsymbol{\lambda}(\tau_j)\cdot \mathrm{f}(\mathrm{x}^j, u^b, e^*(x^j)) \geqslant 0 && (3.38)\\ &\boldsymbol{\lambda}(\tau_j)\cdot \mathrm{f}(\mathrm{x}^j, p^*(x^j), v^b) \leqslant 0 && (3.39)\\ &\boldsymbol{\lambda}(\tau_j)\cdot \mathrm{f}(\mathrm{x}^j, p^*(x^j), e^*(x^j)) = 0 && (3.40)\end{aligned}\right\} \quad \text{where} \quad \mathrm{x}^j = \mathbf{x}^*(\tau_j)$$

$$\forall \tau_j \in [0, \tau_f), \qquad \forall u^b \in K_u(x^j), \qquad \forall v^b \in K_v(x^j)$$

Let $\mathbf{x}^*\colon\ \tau \to x = \mathbf{x}^*(\tau),\ \tau \in [0, \tau_f]$, denote the function that represents π^*.†

Since f is of class C^1 on $G \times U \times V$, and p^* and e^* are of class C^1 on a neighborhood of π^*†, and $\mathbf{x}^*\colon\ \tau_j \to \mathbf{x}^*(\tau_j)$ as well as $\boldsymbol{\lambda}\colon\ \tau_j \to \boldsymbol{\lambda}(\tau_j)$ are continuous functions of τ_j on $[0, \tau_f]$, it follows that

$$H\colon\quad \tau_j \to H(\tau_j)$$

where

$$H(\tau_j) = \boldsymbol{\lambda}(\tau_j) \cdot \mathsf{f}(\mathbf{x}^*(\tau_j), p^*(\mathbf{x}^*(\tau_j)), e^*(\mathbf{x}^*(\tau_j)))$$

is a continuous function of τ_j on $[0, \tau_f]$. Hence (3.40) holds at $\tau_j = \tau_f$ and $\mathsf{x}^j = \mathsf{x}^f$.

One can prove that (3.38) and (3.39) also hold at $\tau_j = \tau_f$. From (3.38) and (3.39) it follows that

$$\boldsymbol{\lambda}(\tau_j) \cdot \mathsf{f}(\mathbf{x}^*(\tau_j), \alpha_{x^f}(\mathbf{x}^*(\tau_j)), e^*(\mathbf{x}^*(\tau_j))) \geqslant 0 \tag{3.41}$$

$$\boldsymbol{\lambda}(\tau_j) \cdot \mathsf{f}(\mathbf{x}^*(\tau_j), p^*(\mathbf{x}^*(\tau_j)), \beta_{x^f}(\mathbf{x}^*(\tau_j))) \leqslant 0 \tag{3.42}$$

where α_{x^f} and β_{x^f} are strategies that satisfy Assumption 3.6 at point $x^f = \mathbf{x}^*(\tau_f)$, with $\alpha_{x^f}(x^f) = u^b$ and $\beta_{x^f}(x^f) = v^b$.

Then allowing $\tau_j \to \tau_f$, we obtain by continuity

$$\boldsymbol{\lambda}(\tau_f) \cdot \mathsf{f}(\mathbf{x}^*(\tau_f), u^b, e^*(\mathbf{x}^*(\tau_f))) \geqslant 0$$

$$\boldsymbol{\lambda}(\tau_f) \cdot \mathsf{f}(\mathbf{x}^*(\tau_f), p^*(\mathbf{x}^*(\tau_f)), v^b) \leqslant 0$$

Clearly, these inequalities are satisfied for all $u^b \in K_u(x^f)$ and all $v^b \in K_v(x^f)$. Hence, conditions (3.38), (3.39) and (3.40) hold for all $\tau_j \in [0, \tau_f]$, for all $u^b \in K_u(x^j)$ and for all $v^b \in K_v(x^j)$. These relations can be rewritten as

$$\min_{u \in K_u(x)} \boldsymbol{\lambda}(\tau) \cdot \mathsf{f}(\mathsf{x}, u, e^*(x)) = \max_{v \in K_v(x)} \boldsymbol{\lambda}(\tau) \cdot \mathsf{f}(\mathsf{x}, p^*(x), v)$$

$$= \boldsymbol{\lambda}(\tau) \cdot \mathsf{f}(\mathsf{x}, p^*(x), e^*(x)), \tag{3.43}$$

$$\boldsymbol{\lambda}(\tau) \cdot \mathsf{f}(\mathsf{x}, p^*(x), e^*(x)) = 0 \qquad \forall \tau \in [0, \tau_f] \tag{3.44}$$

where $\mathsf{x} = \mathbf{x}^*(\tau)$ and $x = \mathbf{x}^*(\tau)$.

Since the functions f_ν, $\nu = 0, 1, \ldots, n$, and the strategies p^* and e^* do not contain x_0 in their arguments, the right-hand side of the adjoint equation which is associated with the zeroth component of λ is zero. It

† Recall that π^* is the projection of $\Pi^*(C)$ on X^*.

follows that $d\lambda_0/d\tau \equiv 0$, and hence

$$\boldsymbol{\lambda}_0(\tau) \equiv \text{constant} \tag{3.45}$$

on $[0, \tau_f)$.

Furthermore, since λ^i is codirectional with $\mathbf{n}(\mathbf{x}^i)$, and since $\mathbf{n}_0(\mathbf{x}^i) = 1$,

$$\boldsymbol{\lambda}_0(\tau) \equiv \text{constant} > 0 \tag{3.46}$$

on $[0, \tau_f)$. Since $\boldsymbol{\lambda}$: $\tau \to \lambda = \boldsymbol{\lambda}(\tau)$ is continuous on $[0, \tau_f]$, (3.46) holds for all $\tau \in [0, \tau_f]$.

3.12 TRANSVERSALITY CONDITION

We shall now investigate conditions which must hold at the point where optimal path $\Pi^*(C)$ reaches Θ.

The intersection of game surface $\Sigma(C)$ with Θ is the set

$$\Theta \cap \Sigma(C) = \{\mathbf{x}\colon \ \Phi(\mathbf{x}) = C, \qquad x \in \theta\} \tag{3.47}$$

Because of (3.7), the condition $x \in \theta$ implies that $V^*(x) = 0$ which, in turn, implies that

$$\Phi(\mathbf{x}) \triangleq x_0 + V^*(x) = x_0$$

Hence,

$$\Theta \cap \Sigma(C) = \{\mathbf{x}\colon \ x_0 = C, \qquad x \in \theta\} \tag{3.48}$$

That is, the intersection of $\Sigma(C)$ with Θ coincides with the intersection of Θ and the plane P_0, perpendicular to x_0-axis, defined by $x_0 = C$. In other words, $\Theta \cap \Sigma(C)$ may be deduced from θ by translation parallel to the x_0-axis.

Now let us consider again the optimal path $\Pi^*(C)$, represented by $\mathbf{x}^*$: $\tau \to \mathbf{x}^*(\tau)$, $\tau \in [0, \tau_f]$, that reaches $\Theta \cap \Sigma(C)$ at terminal point x^f.

Since $\boldsymbol{\lambda}(\tau)$ is codirectional with grad $\Phi(\mathbf{x}^*(\tau))$ for all $\tau \in [0, \tau_f)$, and since $\boldsymbol{\lambda}(\tau)$ and grad $\Phi(\mathbf{x}^*(\tau))$ possess zeroth components $\boldsymbol{\lambda}_0(\tau)$ and 1, respectively,

$$\boldsymbol{\lambda}(\tau) = \boldsymbol{\lambda}_0(\tau) \operatorname{grad} \Phi(\mathbf{x}^*(\tau)) \qquad \forall \tau \in [0, \tau_f)$$

Since the adjoint equation is homogeneous of degree one in λ, and since $\boldsymbol{\lambda}_0(\tau) \equiv \text{constant} > 0$, we may set $\boldsymbol{\lambda}_0(\tau) \equiv 1$. Hence,

$$\boldsymbol{\lambda}(\tau) = \operatorname{grad} \Phi(\mathbf{x}^*(\tau)) \qquad \forall \tau \in [0, \tau_f)$$

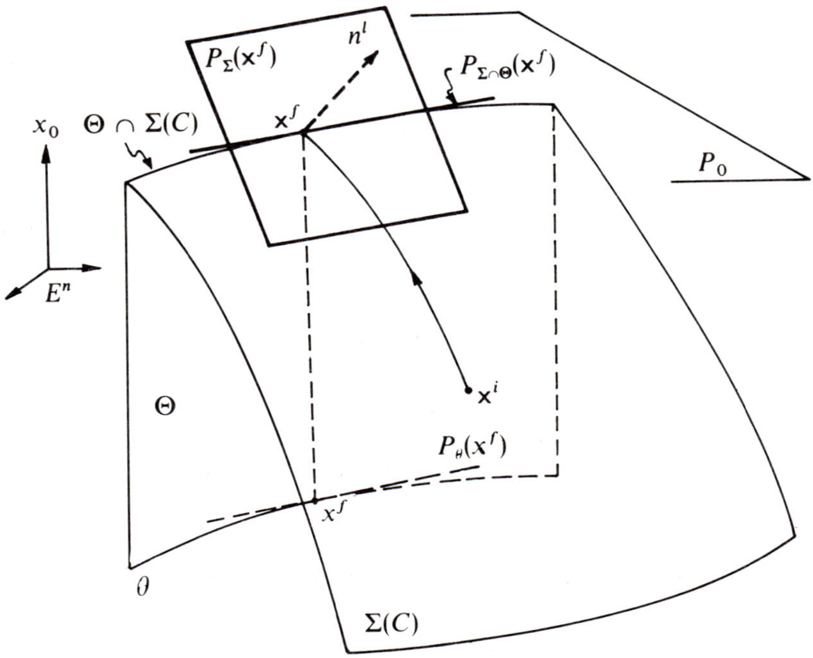

FIG. 3.6. Transversality condition.

Finally, since $\boldsymbol{\lambda}(\tau)$ is continuous on $[0, \tau_f]$, grad $\Phi(\mathbf{x}^*(\tau))$ tends to a limit n^l as $\tau \to \tau_f$, and

$$\boldsymbol{\lambda}(\tau_f) = n^l$$

Thus $\Sigma(C)$ possesses a well defined tangent plane, $P_\Sigma(\mathbf{x}^f)$, at point $\mathbf{x}^f$, namely the plane through $\mathbf{x}^f$ which is perpendicular to n^l.† Since, according to Assumption 3.2, $\partial\theta$ possesses a unique $(n-1)$-dimensional tangent plane, $P_\theta(x)$, at every point x of $\partial\theta$, so that the tangent plane, $P_{\Sigma\cap\Theta}(\mathbf{x})$, of $\partial\Theta \cap \Sigma(C)$ is defined for all $\mathbf{x} \in \partial\Theta \cap \Sigma(C)$,

$$P_{\Sigma\cap\Theta}(\mathbf{x}^f) \subset P_\Sigma(\mathbf{x}^f)$$

Since $\boldsymbol{\lambda}(\tau_f) = n^l$, $\boldsymbol{\lambda}(\tau_f)$ is normal to $P_\Sigma(\mathbf{x}^f)$, and hence to $P_{\Sigma\cap\Theta}(\mathbf{x}^f)$. This orthogonality condition is called the *transversality condition.* It

† This is a consequence of the uniqueness of optimal path $\Pi^*(C)$.

can be expressed as follows. Let $\eta = (\eta_0, \eta_1, \ldots, \eta_n)$ be a vector in $P_{\Sigma\cap\Theta}(\mathrm{x}^f)$; then

$$\boldsymbol{\lambda}(\tau_f) \cdot \eta = 0 \tag{3.49}$$

Since $x_0 = C$ and $x \in P_\theta(x^f)$ for points of $P_{\Sigma\cap\Theta}(\mathrm{x}^f)$, and since $\partial\theta$ is defined by the equation $\boldsymbol{\theta}(x) = 0$, it follows that

$$\begin{aligned} \eta_0 &= 0 \\ \sum_{\nu=1}^{n} \frac{\partial \boldsymbol{\theta}(x)}{\partial x_\nu}\bigg|_{x=x_f} \eta_\nu &= 0 \end{aligned} \tag{3.50}$$

3.13 A MIN-MAX PRINCIPLE, REGULAR CASE

Summarizing the results of the preceding sections, we arrive at a theorem of fundamental importance for the discussion of optimal strategies in differential quantitative games. Letting

$$\mathscr{H}(\lambda, \mathrm{x}, u, v) \triangleq \lambda \cdot \mathrm{f}(\mathrm{x}, u, v)$$

we have

Theorem 3.1. *If* $\Pi^*(C)$ *is an optimal path, generated by strategies* p^* *and* e^*, *represented by* $\mathbf{x}^*\colon\ \tau \to \mathrm{x} = \mathbf{x}^*(\tau),\ \tau \in [0, \tau_f],\ \tau_f \neq 0,$ *and satisfying Assumptions 3.5–3.8, then there exists a nonzero continuous solution* $\boldsymbol{\lambda}\colon\ \tau \to \lambda = \boldsymbol{\lambda}(\tau),\ \tau \in [0, \tau_f]$, *of adjoint equation (3.14) such that*

$$\left.\begin{aligned}
&\text{(a)} \quad \min_{u \in K_u(\mathbf{x}^*(\tau))} \mathscr{H}(\boldsymbol{\lambda}(\tau), \mathbf{x}^*(\tau), u, e^*(\mathbf{x}^*(\tau)) \\
&\qquad = \max_{v \in K_v(\mathbf{x}^*(\tau))} \mathscr{H}(\boldsymbol{\lambda}(\tau), \mathbf{x}^*(\tau), p^*(\mathbf{x}^*(\tau)), v) \\
&\qquad = \mathscr{H}(\boldsymbol{\lambda}(\tau), \mathbf{x}^*(\tau), p^*(\mathbf{x}^*(\tau)), e^*(\mathbf{x}^*(\tau))) \\
&\text{(b)} \quad \mathscr{H}(\boldsymbol{\lambda}(\tau), \mathbf{x}^*(\tau), p^*(\mathbf{x}^*(\tau)), e^*(\mathbf{x}^*(\tau))) = 0 \\
&\text{(c)} \quad \boldsymbol{\lambda}_0(\tau) \equiv \text{constant} > 0
\end{aligned}\right\} \text{for all } \tau \in [0, \tau_f]$$

where $\mathbf{x}^*(\tau)$ *is the projection of* $\mathbf{x}^*(\tau)$ *on* X^*. *Furthermore, if Assumption 3.2 is satisfied, then*

(d) *vector* $(\boldsymbol{\lambda}_1(\tau), \boldsymbol{\lambda}_2(\tau), \ldots, \boldsymbol{\lambda}_n(\tau))$ *is normal to the terminal manifold boundary* $\partial\theta$ *at* $\tau = \tau_f$.

Note that state equation (3.12) and adjoint equation (3.14) may be written as

$$\frac{dx_\nu}{d\tau} = \frac{\partial\mathscr{H}(\lambda, \mathrm{x}, u, v)}{\partial\lambda_\nu}$$

$$\frac{d\lambda_\nu}{d\tau} = -\frac{\partial\mathscr{H}(\lambda, \mathrm{x}, u, v)}{\partial x_\nu} - \sum_{\alpha=1}^{r}\frac{\partial\mathscr{H}(\lambda, \mathrm{x}, u, v)}{\partial u_\alpha}\frac{\partial p_\alpha^*(x)}{\partial x_\nu} - \sum_{\alpha=1}^{s}\frac{\partial\mathscr{H}(\lambda, \mathrm{x}, u, v)}{\partial v_\alpha}\frac{\partial e_\alpha^*(x)}{\partial x_\nu} \qquad \nu = 0, 1, \ldots, n \tag{3.51}$$

where $u = p^*(x)$, $v = e^*(x)$, $\mathrm{x} = \mathbf{x}^*(\tau)$ and $\lambda = \boldsymbol{\lambda}(\tau)$ for $\tau \in [0, \tau_f]$.

By using (3.49) in (a) and (b) of Theorem 3.1 we obtain the functional equation of dynamic programming:

$$\begin{aligned}&\min_{u\in K_u(x)}\left[f_0(\mathrm{x}, u, e^*(x)) + \sum_{\nu=1}^{n} f_\nu(\mathrm{x}, u, e^*(x))\frac{\partial V^*(x)}{\partial x_\nu}\right]_{\mathrm{x}=\mathbf{x}^*(\tau)}\\ &= \max_{v\in K_v(x)}\left[f_0(\mathrm{x}, p^*(x), v) + \sum_{\nu=1}^{n} f_\nu(\mathrm{x}, p^*(x), v)\frac{\partial V^*(x)}{\partial x_\nu}\right]_{\mathrm{x}=\mathbf{x}^*(\tau)} = 0\end{aligned} \tag{3.52}$$

3.14 DISCONTINUITY MANIFOLDS

Now let us suppose that X is domain, and consider

Definition 3.6. We shall say that $\{X_1, X_2, \ldots, X_K\}$ constitutes a *decomposition of* X, if

(i) X_σ, $\sigma = 1, 2, \ldots, K < \infty$, is a domain of E^n, and $\bar{X}_\sigma \subset G$;
(ii) $X_k \cap X_l = \varnothing \qquad k \neq l$
(iii) $X_\sigma \subseteq X \qquad \sigma = 1, 2, \ldots, K$
(iv) $X \subset \bigcup_{\sigma=1}^{K} \bar{X}_\sigma$

As a direct consequence of these conditions, $\bar{X} = \bigcup_{\sigma=1}^{K} \bar{X}_\sigma$.

Next we introduce an assumption that is weaker than Assumption 3.7, namely,

Assumption 3.9. X is a domain possessing a decomposition $\{X_1, X_2, \ldots, X_K\}$ such that functions p^*, e^* agree on each X_σ with functions—say p^σ, e^σ—of class C^1 on a domain $R_\sigma \supset \bar{X}_\sigma$, and on each nonempty $M_{kl} \triangleq \bar{X}_k \cap \bar{X}_l$, $k \neq l$, with functions—say p^{kl}, e^{kl}—of class C^1 on a domain $R_{kl} \supset M_{kl}$.

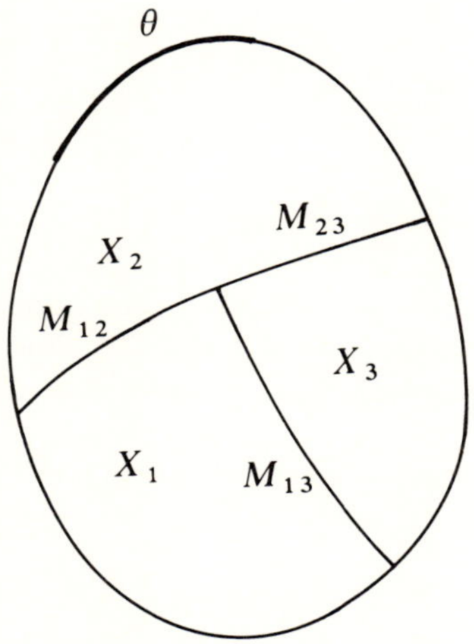

FIG. 3.7. A decomposition of X; discontinuity manifolds in E^n.

Furthermore, we shall make

Assumption 3.10. A *discontinuity manifold* M_{kl} is an (n-1)-dimensional surface which can be represented by

$$\mathbf{m}(x) = 0$$

where function $\mathbf{m}$ is of class C^1 and grad $\mathbf{m}(x) \neq 0$ on a domain containing M_{kl}.

3.15 REGULAR PORTION OF AN OPTIMAL PATH

Definition 3.7. $\{\Sigma_1(C), \Sigma_2(C), \ldots, \Sigma_K(C)\}$ is a *decomposition of* $\Sigma(C)$, where

$$\Sigma_\sigma(C) \triangleq \{\mathrm{x}\colon \ \Phi(\mathrm{x}) \triangleq x_0 + V^*(x) = C, \quad x \in X_\sigma\} \qquad \sigma = 1, 2, \ldots, K$$

And now we make an assumption that is weaker than Assumption 3.8, namely

Assumption 3.11. Function V^* is continuous on X and agrees on each X_σ, $\sigma = 1, 2, \ldots, K$, with a function—say V^σ—twice continuously differentiable on R_σ.

As a consequence of Assumption 3.11,

$$\text{grad}\,\Phi(\mathsf{x}) = \left(1, \frac{\partial V^*(x)}{\partial x_1}, \ldots, \frac{\partial V^*(x)}{\partial x_n}\right) \tag{3.53}$$

is defined for all $\mathsf{x} \in \mathscr{X}_\sigma \triangleq X_\sigma \times \{x_0\}$, and all points of $\Sigma_\sigma(C)$,

$$\sigma = 1, 2, \ldots, K,$$

are regular interior points of $\Sigma_\sigma(C)$.

Here we shall relax the assumptions of Theorem 3.1; that is, we shall be concerned with an optimal path $\Pi^*(C)$, generated by strategy pair (p^*, e^*), satisfying Assumptions 3.5 and 3.6, and we shall derive conditions (a)–(c) of Theorem 3.1 for points of a regular portion of $\Pi^*(C)$, namely $\Pi^*(C) \cap \Sigma_\sigma(C)$.

By the same arguments as those employed in Section 3.8, it follows from Lemmas 2.1 and 2.2, and Corollary 2.1, that

$$\text{grad}\,\Phi(\mathsf{x}) \cdot \mathsf{f}(\mathsf{x}, p^*(x), v) \leqslant 0 \tag{3.54}$$

$$\text{grad}\,\Phi(\mathsf{x}) \cdot \mathsf{f}(\mathsf{x}, u, e^*(x)) \geqslant 0 \tag{3.55}$$

$$\text{grad}\,\Phi(\mathsf{x}) \cdot \mathsf{f}(\mathsf{x}, p^*(x), e^*(x)) = 0 \tag{3.56}$$

for all $\mathsf{x} \in \Pi^*(C) \cap \Sigma_\sigma(C)$, for all $u \in K_u(x)$ and for all $v \in K_v(x)$.

Since (p^*, e^*) is optimal on X, (3.56) holds for all $\mathsf{x} \in \Sigma_\sigma(C)$,

$$\sigma = 1, 2, \ldots, K,$$

and for all C; that is, (3.56) is an identity in x on $\mathscr{X}_\sigma$, $\sigma = 1, 2, \ldots, K$.

Let us rewrite (3.56) as

$$\sum_{\alpha=0}^{n} \frac{\partial \Phi(\mathsf{x})}{\partial x_\alpha} f_\alpha(\mathsf{x}, p^*(x), e^*(x)) = 0 \tag{3.57}$$

By differentiating (3.57) with respect to x_β, $\beta = 0, 1, \ldots, n$, we obtain

$$\sum_{\alpha=0}^{n} \frac{\partial^2 \Phi(\mathsf{x})}{\partial x_\alpha\, \partial x_\beta} f_\alpha(\mathsf{x}, p^*(x), e^*(x)) + \sum_{\alpha=0}^{n} \frac{\partial \Phi(\mathsf{x})}{\partial x_\alpha} \frac{\partial f_\alpha(\mathsf{x}, p^*(x), e^*(x))}{\partial x_\beta} = 0, \qquad \beta = 0, 1, \ldots, n \tag{3.58}$$

Now, along a portion of optimal path $\Pi^*(C)$ lying on $\Sigma_\sigma(C)$, we have

$$\frac{d}{d\tau}\left(\frac{\partial \Phi(\mathsf{x})}{\partial x_\beta}\right) = \sum_{\alpha=0}^{n} \frac{\partial^2 \Phi(\mathsf{x})}{\partial x_\alpha\, \partial x_\beta} f_\alpha(\mathsf{x}, p^*(x), e^*(x)), \qquad \beta = 0, 1, \ldots, n \quad (3.59)$$

From (3.58) and (3.59) we obtain

$$\frac{d}{d\tau}\left(\frac{\partial \Phi(\mathsf{x})}{\partial x_\beta}\right) = -\sum_{\alpha=0}^{n} \frac{\partial \Phi(\mathsf{x})}{\partial x_\alpha} \frac{\partial f_\alpha(\mathsf{x}, p^*(x), e^*(x))}{\partial x_\beta}, \qquad \beta = 0, 1, \ldots, n \quad (3.60)$$

By developing $\partial f_\alpha(\mathsf{x}, p^*(x), e^*(x))/\partial x_\beta$, and writing (3.60) in vector form, we obtain

$$\frac{d}{d\tau} \operatorname{grad} \Phi(\mathsf{x}) = -\left(\frac{\partial \mathsf{f}}{\partial \mathsf{x}} + \frac{\partial \mathsf{f}}{\partial u}\frac{\partial p^*}{\partial \mathsf{x}} + \frac{\partial \mathsf{f}}{\partial v}\frac{\partial e^*}{\partial \mathsf{x}}\right)^{\mathrm{T}} \operatorname{grad} \Phi(\mathsf{x}) \quad (3.61)$$

for $\mathsf{x} = \mathbf{x}^*(\tau) \in \Sigma_\sigma(C)$.

Letting

$$\boldsymbol{\lambda}(\tau) \triangleq \operatorname{grad} \Phi(\mathsf{x}), \qquad \mathsf{x} = \mathbf{x}^*(\tau) \in \Sigma_\sigma(C) \quad (3.62)$$

and substituting in (3.54), (3.55) and (3.56), we obtain

$$\mathscr{H}(\boldsymbol{\lambda}(\tau), \mathbf{x}^*(\tau), p^*(\mathbf{x}^*(\tau)), v) \leqslant 0 \quad (3.63)$$

$$\mathscr{H}(\boldsymbol{\lambda}(\tau), \mathbf{x}^*(\tau), u, e^*(\mathbf{x}^*(\tau))) \geqslant 0 \quad (3.64)$$

$$\mathscr{H}(\boldsymbol{\lambda}(\tau), \mathbf{x}^*(\tau), p^*(\mathbf{x}^*(\tau)), e^*(\mathbf{x}^*(\tau))) = 0 \quad (3.65)$$

Of course, conditions (3.63) and (3.64) must hold for all $u \in K_u(\mathbf{x}^*(\tau))$ and for all $v \in K_v(\mathbf{x}^*(\tau))$.

Furthermore, it follows from (3.53) and (3.62) that

$$\boldsymbol{\lambda}_0(\tau) \equiv 1 \quad (3.66)$$

Clearly, conditions (3.63)–(3.66) are identical with conditions (a)–(c) of Theorem 3.1; that is, the above method is another way of deriving Theorem 3.1, since it obviously applies to regular optimal paths on $\Sigma(C)$.† We shall let the reader verify that relations (3.63)–(3.66) can also be obtained by invoking arguments similar to those of the previous sections; that is, by considering transfer of the tangent plane of $\Sigma(C)$ along a portion of $\Pi^*(C)$ lying on $\Sigma_\sigma(C)$.

† However, here we require that $\Phi(\mathsf{x})$ is twice continuously differentiable.

3.16 JUMP CONDITION

Definition 3.8. We shall say that optimal path $\Pi^*(C)$ *crosses* $\mathscr{M}_{kl} \triangleq M_{kl} \times \{x_0\}$ at point $x^c = \mathbf{x}^*(\tau_c)$, if there exists $\epsilon > 0$ such that

$$\mathbf{x}^*(\tau) \in \Sigma_k(C) \quad \text{for all} \quad \tau \in (\tau_c - \epsilon, \tau_c); \quad \text{and}$$

$$\mathbf{x}^*(\tau) \in \Sigma_l(C) \quad \text{for all} \quad \tau \in (\tau_c, \tau_c + \epsilon); \quad \text{and}$$

$$\text{grad } \mathbf{m}(x^c) \cdot f(x^c, p^k(x^c), e^k(x^c)) \neq 0$$

$$\text{grad } \mathbf{m}(x^c) \cdot f(x^c, p^l(x^c), e^l(x^c)) \neq 0$$

We shall now investigate conditions which must hold at a point $\mathrm{x}^c = \mathbf{x}^*(\tau_c)$ where optimal path $\Pi^*(C)$ crosses $\mathscr{M}_{kl}$.

Since p^*, e^* agree on X_k with functions p^k, e^k, respectively, which are of class C^1 on a domain $R_k \supset \bar{X}_k$, and since $\boldsymbol{\lambda}$: $\tau \to \lambda = \boldsymbol{\lambda}(\tau)$,

$$\tau \in (\tau_c - \epsilon, \tau_c),$$

is a solution of

$$\frac{d\lambda}{d\tau} = -\left(\frac{\partial \mathrm{f}}{\partial \mathrm{x}} + \frac{\partial \mathrm{f}}{\partial u}\frac{\partial p^k}{\partial \mathrm{x}} + \frac{\partial \mathrm{f}}{\partial v}\frac{\partial e^k}{\partial \mathrm{x}}\right)^{\mathrm{T}}\Bigg|_{\mathrm{x}=\mathbf{x}^*(\tau)} \lambda$$

$\boldsymbol{\lambda}(\tau)$ tends to a limit $\boldsymbol{\lambda}(\tau_c - 0)$ as $\tau \to \tau_c$, $\tau < \tau_c$. Since $\boldsymbol{\lambda}(\tau) = \text{grad } \Phi(\mathbf{x}^*(\tau))$ for all $\tau \in (\tau_c - \epsilon, \tau_c)$, $\lim_{\tau \to \tau_c, \tau < \tau_c} \text{grad } \Phi(\mathbf{x}^*(\tau))$ is defined.

By similar arguments, based on the fact that p^*, e^* agree on X_l with functions p^l, e^l, respectively, which are of class C^1 on a domain $R_l \supset \bar{X}_l$, one can prove that $\lim_{\tau \to \tau_c, \tau > \tau_c} \text{grad } \Phi(\mathbf{x}^*(\tau))$ is also defined.

Let

$$\Phi^\sigma(\mathrm{x}) = x_0 + V^\sigma(x), \quad x \in R_\sigma, \quad \sigma = k, l$$

Since

$$\text{grad } \Phi(\mathbf{x}^*(\tau)) = \text{grad } \Phi^k(\mathbf{x}^*(\tau)) \quad \text{for all} \quad \tau \in (\tau_c - \epsilon, \tau_c)$$

$$\text{grad } \Phi(\mathbf{x}^*(\tau)) = \text{grad } \Phi^l(\mathbf{x}^*(\tau)) \quad \text{for all} \quad \tau \in (\tau_c, \tau_c + \epsilon)$$

and since grad $\Phi^k(\mathrm{x}^c)$ and grad $\Phi^l(\mathrm{x}^c)$ are defined, we have

$$\lim_{\substack{\tau \to \tau_c \\ \tau < \tau_c}} \text{grad } \Phi(\mathbf{x}^*(\tau)) = \text{grad } \Phi^k(\mathrm{x}^c)$$

$$\lim_{\substack{\tau \to \tau_c \\ \tau > \tau_c}} \text{grad } \Phi(\mathbf{x}^*(\tau)) = \text{grad } \Phi^l(\mathrm{x}^c)$$

Letting

$$\operatorname{grad}\Phi(\mathbf{x}^c)_- \triangleq \lim_{\substack{\tau\to\tau_c\\ \tau<\tau_c}} \operatorname{grad}\Phi(\mathbf{x}^*(\tau))$$

$$\operatorname{grad}\Phi(\mathbf{x}^c)_+ \triangleq \lim_{\substack{\tau\to\tau_c\\ \tau>\tau_c}} \operatorname{grad}\Phi(\mathbf{x}^*(\tau))$$

We have

$$\operatorname{grad}\Phi(\mathbf{x}^c)_- = \operatorname{grad}\Phi^k(\mathbf{x}^c) = \boldsymbol{\lambda}(\tau_c - 0)$$

$$\operatorname{grad}\Phi(\mathbf{x}^c)_+ = \operatorname{grad}\Phi^l(\mathbf{x}^c) = \boldsymbol{\lambda}(\tau_c + 0)$$

Thus, $\bar{\Sigma}_k(C)$ and $\bar{\Sigma}_l(C)$ possess well defined tangent planes, $P_{\bar{\Sigma}_k}(\mathbf{x}^c)$ and $P_{\bar{\Sigma}_l}(\mathbf{x}^c)$, respectively, at point $\mathbf{x}^c$, namely the planes through $\mathbf{x}^c$ which are perpendicular to $\boldsymbol{\lambda}(\tau_c - 0)$ and $\boldsymbol{\lambda}(\tau_c + 0)$, respectively.

Moreover, according to Assumption 3.10., M_{kl} possesses a unique (n-1)-dimensional tangent plane, $P_M(x)$, at every point x of M_{kl}, and hence the tangent plane $P_{\Sigma\cap\mathscr{M}}(\mathbf{x}^c)$ of $\mathscr{M}_{kl} \cap \Sigma(C)$ is defined at point $\mathbf{x}^c$. And so we conclude that

$$P_{\Sigma\cap\mathscr{M}}(x^c) \subseteq P_{\bar{\Sigma}_k}(\mathbf{x}^c) \cap P_{\bar{\Sigma}_l}(\mathbf{x}^c)$$

Since $\boldsymbol{\lambda}(\tau_c - 0)$, $\boldsymbol{\lambda}(\tau_c + 0)$ and $\operatorname{grad}\mathbf{m}(\mathbf{x}^c)$† are normal to the ($n$-1)-dimensional plane $P_{\Sigma\cap\mathscr{M}}(\mathbf{x}^c)$, these vectors are linearly dependent; that is, there exist constants c_1, c_2, c_3, not all of which are zero, such that

$$c_1\boldsymbol{\lambda}(\tau_c - 0) + c_2\boldsymbol{\lambda}(\tau_c + 0) + c_3 \operatorname{grad}\mathbf{m}(\mathbf{x}^c) = 0 \tag{3.67}$$

In view of (3.66),

$$\boldsymbol{\lambda}_0(\tau_c - 0) = \boldsymbol{\lambda}_0(\tau_c + 0) = 1$$

Furthermore, since the zeroth component of $\operatorname{grad}\mathbf{m}(\mathbf{x}^c)$ is zero, we have $c_1 = -c_2$; and since $\operatorname{grad}\mathbf{m}(\mathbf{x}^c) \neq 0$,

$$c_1 = -c_2 \neq 0$$

Thus, (3.67) can be rewritten as

$$\boldsymbol{\lambda}(\tau_c - 0) - \boldsymbol{\lambda}(\tau_c + 0) + (c_3/c_1) \operatorname{grad}\mathbf{m}(\mathbf{x}^c) = 0 \tag{3.68}$$

Equation (3.68) is the *jump condition* at a point where optimal path $\Pi^*(C)$ crosses discontinuity manifold $\mathscr{M}_{kl}$.

† Of course, $\mathbf{m}(\mathbf{x}^c) = \mathbf{m}(x^c)$.

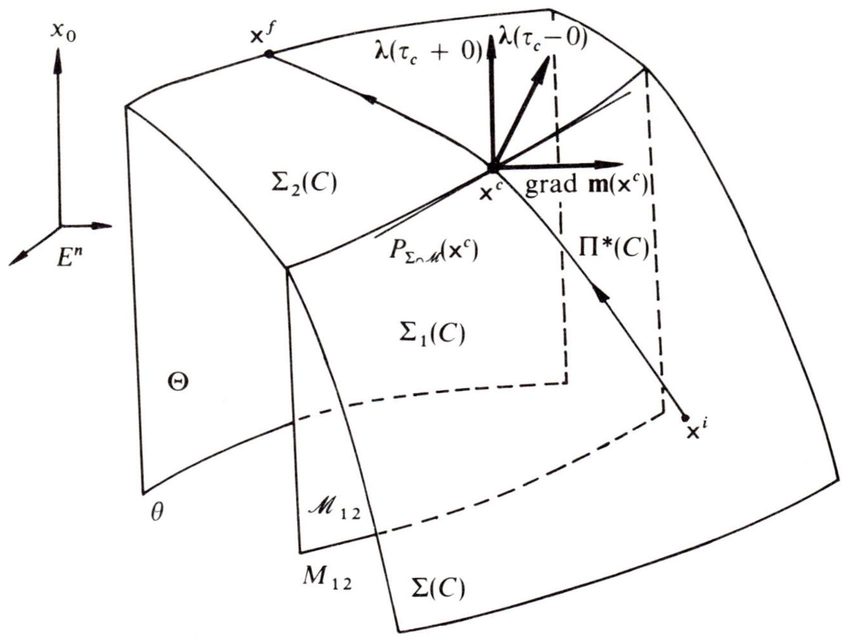

FIG. 3.8. Optimal path crossing a discontinuity manifold in E^{n+1}; jump condition.

Letting

$$\mathsf{f}^- = (f_0^-, f_1^-, \ldots, f_n^-) \triangleq \lim_{\substack{\tau\to\tau_c \\ \tau<\tau_c}} \mathsf{f}(\mathbf{x}^*(\tau), p^*(\mathbf{x}^*(\tau)), e^*(\mathbf{x}^*(\tau)))$$

$$\mathsf{f}^+ = (f_0^+, f_1^+, \ldots, f_n^+) \triangleq \lim_{\substack{\tau\to\tau_c \\ \tau>\tau_c}} \mathsf{f}(\mathbf{x}^*(\tau), p^*(\mathbf{x}^*(\tau)), e^*(\mathbf{x}(\tau)))$$

we deduce from (3.65) and (3.68) that

$$\frac{c_3}{c_1} = -\frac{f_0^+ + \sum_{\nu=1}^{n} \boldsymbol{\lambda}_\nu(\tau_c - 0) f_\nu^+}{\sum_{\nu=1}^{n} \dfrac{\partial \mathbf{m}}{\partial x_\nu} f_\nu^+} = \frac{f_0^- + \sum_{\nu=1}^{n} \boldsymbol{\lambda}_\nu(\tau_c + 0) f_\nu^-}{\sum_{\nu=1}^{n} \dfrac{\partial \mathbf{m}}{\partial x_\nu} f_\nu^-} \tag{3.69}$$

where partial derivatives $\partial\mathbf{m}/\partial x_\nu$, $\nu = 1, 2, \ldots, n$, are evaluated at $x = x^c$.

3.17 CONSTRAINTS

Let us consider constraint sets $K_u(x)$ and $K_v(x)$, respectively, given by

$$\begin{aligned} \varphi_\alpha(x, u) &\leqslant 0 \qquad \alpha = 1, 2, \ldots, k \\ \psi_\alpha(x, v) &\leqslant 0 \qquad \alpha = 1, 2, \ldots, l \end{aligned} \tag{3.70}$$

where functions φ_α and ψ_α are of class C^1 on $G \times U$ and $G \times V$, respectively.

At a point $x = \mathbf{x}^*(\tau)$, $\tau \in [0, \tau_f]$, of the projection on X^* of an optimal path $\Pi^*(C)$, some or all of inequality constraints (3.70) may be equalities. Suppose that

$$\begin{aligned} \varphi_\alpha(x, p^*(x)) &= 0 \qquad \alpha = 1, 2, \ldots, k' \leqslant k \\ \psi_\alpha(x, e^*(x)) &= 0 \qquad \alpha = 1, 2, \ldots, l' \leqslant l \end{aligned} \tag{3.71}$$

We shall suppose also that if $k > r$ or $l > s$, at most r of the $\varphi_\alpha(x, u)$ or s of the $\psi_\alpha(x, v)$ vanish at any point of $G \times U$ and $G \times V$, respectively, and furthermore that the matrices

$$\left[\frac{\partial \varphi_\alpha(x, u)}{\partial u_\sigma}\right] \qquad \begin{aligned} \alpha &= 1, 2, \ldots, k' \\ \sigma &= 1, 2, \ldots, r \end{aligned}$$

$$\left[\frac{\partial \psi_\alpha(x, v)}{\partial v_\sigma}\right] \qquad \begin{aligned} \alpha &= 1, 2, \ldots, l' \\ \sigma &= 1, 2, \ldots, s \end{aligned}$$

have maximum rank at $x = \mathbf{x}^*(\tau)$, $u = p^*(\mathbf{x}^*(\tau))$, $v = e^*(\mathbf{x}^*(\tau))$.

Now, at a point $\mathbf{x} \in \Pi^*(C)$ where condition (a) of Theorem 3.1 applies, the Lagrange multiplier rule† gives

$$\begin{aligned} \frac{\partial \mathscr{H}}{\partial u} + \mu^{\mathrm{T}} \frac{\partial \varphi}{\partial u} &= \lambda^{\mathrm{T}} \frac{\partial \mathrm{f}}{\partial u} + \mu^{\mathrm{T}} \frac{\partial \varphi}{\partial u} = 0 \\ \frac{\partial \mathscr{H}}{\partial v} + \nu^{\mathrm{T}} \frac{\partial \psi}{\partial v} &= \lambda^{\mathrm{T}} \frac{\partial \mathrm{f}}{\partial v} + \nu^{\mathrm{T}} \frac{\partial \psi}{\partial v} = 0 \end{aligned} \tag{3.72}$$

† Kuhn–Tucker conditions.

where $\partial\mathscr{H}/\partial u$, $\partial\mathscr{H}/\partial v$, λ^{T}, μ^{T}, and ν^{T} are row vectors:

$$\frac{\partial\mathscr{H}}{\partial u} \triangleq \left(\frac{\partial\mathscr{H}}{\partial u_1}\frac{\partial\mathscr{H}}{\partial u_2}\cdots\frac{\partial\mathscr{H}}{\partial u_r}\right)$$
$$\frac{\partial\mathscr{H}}{\partial v} \triangleq \left(\frac{\partial\mathscr{H}}{\partial v_1}\frac{\partial\mathscr{H}}{\partial v_2}\cdots\frac{\partial\mathscr{H}}{\partial v_s}\right)$$
$$\lambda^{\mathrm{T}} \triangleq (\lambda_0\,\lambda_1\,\ldots\,\lambda_n)$$
$$\mu^{\mathrm{T}} \triangleq (\mu_1\,\mu_2\,\ldots\,\mu_{k'}\,0\cdots0) \qquad \text{where } \mu_i \geqslant 0$$
$$\nu^{\mathrm{T}} \triangleq (\nu_1\,\nu_2\,\ldots\,\nu_{l'}\,0\cdots0) \qquad \text{where } \nu_i \leqslant 0$$

and

$$\frac{\partial\varphi}{\partial u} \triangleq \left[\frac{\partial\varphi_\alpha(x,u)}{\partial u_\sigma}\right] \qquad \begin{aligned}\alpha &= 1,2,\ldots,k\\ \sigma &= 1,2,\ldots,r\end{aligned}$$
$$\frac{\partial\psi}{\partial v} \triangleq \left[\frac{\partial\psi_\alpha(x,v)}{\partial v_\sigma}\right] \qquad \begin{aligned}\alpha &= 1,2,\ldots,l\\ \sigma &= 1,2,\ldots,s\end{aligned}$$

All the partial derivatives are evaluated at

$$u = p^*(\mathbf{x}^*(\tau)), \qquad v = e^*(\mathbf{x}^*(\tau)), \qquad \mathsf{x} = \mathbf{x}^*(\tau)$$

From (3.70) and (3.71) we see that $\varphi_\alpha(x, p^*(x))$, $\alpha = 1, 2, \ldots, k'$, and $\psi_\alpha(x, e^*(x))$, $\alpha = 1, 2, \ldots, l'$, have a stationary maximum for

$$x = \mathbf{x}^*(\tau) \notin \theta.$$

Hence we may write†

$$\begin{aligned}\sum_{\sigma=1}^{n}\frac{\partial\varphi_\alpha}{\partial x_\sigma} + \sum_{\sigma=1}^{n}\sum_{\beta=1}^{r}\frac{\partial\varphi_\alpha}{\partial u_\beta}\frac{\partial p_\beta^*}{\partial x_\sigma} &= 0 \qquad \alpha = 1,2,\ldots,k'\\ \sum_{\sigma=1}^{n}\frac{\partial\psi_\alpha}{\partial x_\sigma} + \sum_{\sigma=1}^{n}\sum_{\beta=1}^{l}\frac{\partial\psi_\alpha}{\partial v_\beta}\frac{\partial e_\beta^*}{\partial x_\sigma} &= 0 \qquad \alpha = 1,2,\ldots,l'\end{aligned} \tag{3.73}$$

at $u = p^*(\mathbf{x}^*(\tau))$, $v = e^*(\mathbf{x}^*(\tau))$, $x = \mathbf{x}^*(\tau) \neq \mathbf{x}^*(\tau_f)$

Equations (3.73) may be written in vector form

$$\begin{aligned}\mu^{\mathrm{T}}\left[\frac{\partial\varphi}{\partial\mathsf{x}} + \frac{\partial\varphi}{\partial u}\frac{\partial p^*}{\partial\mathsf{x}}\right] &= 0\\ \nu^{\mathrm{T}}\left[\frac{\partial\psi}{\partial\mathsf{x}} + \frac{\partial\psi}{\partial v}\frac{\partial e^*}{\partial\mathsf{x}}\right] &= 0\end{aligned} \tag{3.74}$$

† This result holds for $\tau \in [0, \tau_f)$, the half open interval since all but the terminal point are assumed to be interior points of X.

where

$$\frac{\partial \varphi}{\partial \mathsf{x}} \triangleq \left[\frac{\partial \varphi_\alpha(x, u)}{\partial x_\sigma}\right] \qquad \begin{matrix} \alpha = 1, 2, \ldots, k \\ \sigma = 0, 1, \ldots, n \end{matrix}$$

$$\frac{\partial \psi}{\partial \mathsf{x}} \triangleq \left[\frac{\partial \psi_\alpha(x, v)}{\partial x_\sigma}\right] \qquad \begin{matrix} \alpha = 1, 2, \ldots, l \\ \sigma = 0, 1, \ldots, n \end{matrix}$$

evaluated at $u = p^*(\mathbf{x}^*(\tau))$, $v = e^*(\mathbf{x}^*(\tau))$, $x = \mathbf{x}^*(\tau)$.

From (3.72) and (3.74) we obtain

$$\lambda^{\mathrm{T}} \frac{\partial \mathsf{f}}{\partial u} \frac{\partial p^*}{\partial \mathsf{x}} = \mu^{\mathrm{T}} \frac{\partial \varphi}{\partial \mathsf{x}}$$

$$\lambda^{\mathrm{T}} \frac{\partial \mathsf{f}}{\partial v} \frac{\partial e^*}{\partial \mathsf{x}} = \nu^{\mathrm{T}} \frac{\partial \psi}{\partial \mathsf{x}}$$

or equivalently

$$\left(\frac{\partial \mathsf{f}}{\partial u} \frac{\partial p^*}{\partial \mathsf{x}}\right)^{\mathrm{T}} \lambda = \left(\frac{\partial \varphi}{\partial \mathsf{x}}\right)^{\mathrm{T}} \mu$$
$$\left(\frac{\partial \mathsf{f}}{\partial v} \frac{\partial e^*}{\partial \mathsf{x}}\right)^{\mathrm{T}} \lambda = \left(\frac{\partial \psi}{\partial \mathsf{x}}\right)^{\mathrm{T}} \nu \qquad (3.75)$$

where μ and ν are column vectors:

$$\mu = \begin{pmatrix} \mu_1 \\ \mu_2 \\ \cdot \\ \cdot \\ \cdot \\ \mu_{k'} \\ 0 \\ \cdot \\ \cdot \\ \cdot \\ 0 \end{pmatrix} \qquad \nu = \begin{pmatrix} \nu_1 \\ \nu_2 \\ \cdot \\ \cdot \\ \cdot \\ \nu_{l'} \\ 0 \\ \cdot \\ \cdot \\ \cdot \\ 0 \end{pmatrix}$$

Finally, in view of (3.75) adjoint equation (3.14) may be written as

$$\frac{d\lambda}{d\tau} = -\left(\frac{\partial \mathsf{f}}{\partial \mathsf{x}}\right)^{\mathrm{T}} \lambda - \left(\frac{\partial \varphi}{\partial \mathsf{x}}\right)^{\mathrm{T}} \mu - \left(\frac{\partial \psi}{\partial \mathsf{x}}\right)^{\mathrm{T}} \nu \qquad (3.76)$$

for $\tau \in [0, \tau_f)$. By continuity of $\partial \mathsf{f}/\partial \mathsf{x}$, $\partial \varphi/\partial \mathsf{x}$, $\partial \psi/\partial \mathsf{x}$ on $[\tau_f - \epsilon, \tau_f]$ for some $\epsilon > 0$, Eq. (3.76) holds on $[0, \tau_f]$.

If constraints (3.70) are independent of x, the adjoint equation reduces to

$$\frac{d\lambda}{d\tau} = -\left(\frac{\partial \mathbf{f}}{\partial \mathbf{x}}\right)^{\mathrm{T}} \lambda \tag{3.77}$$

3.18 A MIN-MAX PRINCIPLE WITH JUMP CONDITION

At last we arrive at

Theorem 3.2. *If there exists a strategy pair* (p^*, e^*) *optimal on* $X \in G$, *and a decomposition* $\{X_1, X_2, \ldots, X_K\}$, $K < \infty$, *of* X *such that Assumptions* 3.9 *and* 3.11 *are satisfied; if* $\Pi^*(C)$ *is an optimal path generated by* (p^*, e^*), *represented by* $\mathbf{x}^*\colon\ \tau \to \mathbf{x} = \mathbf{x}^*(\tau)$, $\tau \in [0, \tau_f]$, *satisfying Assumption* 3.5; *and if the constraint sets are given by* (3.70); *then, on every interval* $[\tau_\alpha, \tau_\beta] \subseteq [0, \tau_f]$ *such that* $\mathbf{x}^*(\tau) \in \mathscr{X}_\sigma \triangleq X_\sigma \times \{x_0\}$,

$$\sigma \in \{1, 2, \ldots, K\},$$

for $\tau \in (\tau_\alpha, \tau_\beta)$, *there exists a nonzero continuous solution* $\boldsymbol{\lambda}\colon\ \tau \to \lambda = \boldsymbol{\lambda}(\tau)$ *of adjoint equation* (3.76), *such that*

$$\left.\begin{aligned}
&\text{(a)}\quad \min_{u \in K_u(\mathbf{x}^*(\tau))} \mathscr{H}(\boldsymbol{\lambda}(\tau), \mathbf{x}^*(\tau), u, e^*(\mathbf{x}^*(\tau)))\\
&\qquad = \max_{v \in K_v(\mathbf{x}^*(\tau))} \mathscr{H}(\boldsymbol{\lambda}(\tau), \mathbf{x}^*(\tau), p^*(\mathbf{x}^*(\tau)), v)\\
&\qquad = \mathscr{H}(\boldsymbol{\lambda}(\tau), \mathbf{x}^*(\tau), p^*(\mathbf{x}^*(\tau)), e^*(\mathbf{x}^*(\tau)))\\
&\text{(b)}\quad \mathscr{H}(\boldsymbol{\lambda}(\tau), \mathbf{x}^*(\tau), p^*(\mathbf{x}^*(\tau)), e^*(\mathbf{x}^*(\tau))) = 0\\
&\text{(c)}\quad \boldsymbol{\lambda}_0(\tau) = \text{constant} > 0
\end{aligned}\right\}\ \text{for all } \tau \in [\tau_\alpha, \tau_\beta]$$

where $\mathbf{x}^*(\tau)$ *is the projection of* $\mathbf{x}^*(\tau)$ *on* X^*.

If Assumption 3.2 *is satisfied, and if there exist* $\sigma \in \{1, 2, \ldots, K\}$ *and* $\epsilon > 0$, *such that* $\mathbf{x}^*(\tau) \in \mathscr{X}_\sigma$ *for all* $\tau \in (\tau_f - \epsilon, \tau_f)$, *then*

(d) *vector* $(\boldsymbol{\lambda}_1(\tau), \boldsymbol{\lambda}_2(\tau), \ldots, \boldsymbol{\lambda}_n(\tau))$ *is normal to the terminal manifold* θ *at* $\tau = \tau_f$.

Furthermore, if $\Pi^*(C)$ *crosses a discontinuity manifold*

$$\mathscr{M}_{kl} \triangleq M_{kl} \times \{x_0\}$$

at $\tau = \tau_c$, *and if Assumption* 3.10 *is satisfied, then*

(e) *jump condition* (3.68) *is satisfied at* $\mathbf{x}^*(\tau_c)$.

Conditions (a)–(c) of Theorem 3.2 are direct consequences of (3.61)–(3.65).

Condition (d) follows from Assumption 3.11 according to which the tangent plane of $\overline{\Sigma}_\sigma(C)$ is well defined at $\mathbf{x}^*(\tau_f)$, together with

$$\lim_{\substack{\tau\to\tau_f\\ \tau<\tau_f}} \operatorname{grad} \Phi\,(\mathbf{x}^*(\tau)) = \boldsymbol{\lambda}(\tau_f)$$

Condition (e) was derived in Sec. 3.16.

3.19 EXAMPLE 3.1

Here we shall consider a game described by

$$\frac{dx_1}{d\tau} = u_1 + v, \qquad \frac{dx_2}{d\tau} = u_2, \qquad \frac{dx_3}{d\tau} = 1 \tag{3.78}$$

with constraints

$$\begin{aligned} K_u&:\quad u_1^2 + u_2^2 \leqslant w^2 \\ K_v&:\quad |v| \leqslant 1 \end{aligned} \tag{3.79}$$

where w is a positive constant, $w < 1$.

Target θ is defined by

$$\theta \triangleq \{x = (x_1, x_2, x_3):\quad x_1^2 + x_2^2 \leqslant R^2\} \tag{3.80}$$

We shall consider the time-optimal problem; namely, J_P seeks to minimize while J_E seeks to maximize the time of transfer from an initial point to the target. In this problem $G = E^3$, and we shall suppose that there exists a strategy pair (p^*, e^*), $p^* \triangleq (p_1^*, p_2^*)$, that is optimal on a subset X^* of G.

As in the previous sections, we shall represent a path in G by a function $\mathbf{x}$: $\tau \to x = \mathbf{x}(\tau)$, and we shall denote by $\mathbf{x}_\alpha(\tau)$, $\alpha = 1, 2, 3$, the components of vector $\mathbf{x}(\tau)$ on orthogonal axes in E^3.

A path in G, generated by a strategy pair that satisfies the necessary conditions for optimality, will be denoted by π^* and represented by $\mathbf{x}^*$: $\tau \to x = \mathbf{x}^*(\tau)$,† although we cannot state that it is optimal without further investigation. This path is only a "candidate" for optimality.

As before, $\boldsymbol{\lambda}$: $\tau \to \lambda = \boldsymbol{\lambda}(\tau)$ will denote a solution of the adjoint equation on some interval, and $\boldsymbol{\lambda}_\alpha(\tau)$, $\alpha = 0, 1, 2, 3$, will denote the

† In Secs. 3.19–3.21 we shall be concerned only with *non-null* paths π^*.

components of vector $\boldsymbol{\lambda}(\tau)$ on orthogonal axes in the augmented state space E^4.

To apply the necessary conditions we have need of the $\mathscr{H}$-function; here

$$\mathscr{H}(\lambda, x, u, v) = 1 + \lambda_1 u_1 + \lambda_2 u_2 + \lambda_1 v + \lambda_3 \tag{3.81}$$

The adjoint equations are

$$d\lambda_\alpha / d\tau = 0 \qquad \alpha = 1, 2, 3 \tag{3.82}$$

so that

$$\boldsymbol{\lambda}_\alpha(\tau) \equiv \text{constant} \qquad \alpha = 1, 2, 3 \tag{3.83}$$

From the transversality condition we obtain

$$\begin{aligned} \boldsymbol{\lambda}_1(\tau_f) &= c\mathbf{x}_1^*(\tau_f) \\ \boldsymbol{\lambda}_2(\tau_f) &= c\mathbf{x}_2^*(\tau_f), \\ \boldsymbol{\lambda}_3(\tau_f) &= 0 \end{aligned} \tag{3.84}$$

where, in view of (3.80),

$$c^2 = [\boldsymbol{\lambda}_1^2(\tau_f) + \boldsymbol{\lambda}_2^2(\tau_f)]/R^2 \tag{3.85}$$

By (3.83) and (3.84) we have

$$\begin{aligned} \boldsymbol{\lambda}_1(\tau) &\equiv c\mathbf{x}_1^*(\tau_f) \\ \boldsymbol{\lambda}_2(\tau) &\equiv c\mathbf{x}_2^*(\tau_f) \\ \boldsymbol{\lambda}_3(\tau) &\equiv 0 \end{aligned} \tag{3.86}$$

From condition (a) of Theorem 3.2 we obtain

$$\begin{aligned} \mathbf{u}_1^*(\tau) &= -w \frac{\boldsymbol{\lambda}_1(\tau)}{\sqrt{\boldsymbol{\lambda}_1^2(\tau) + \boldsymbol{\lambda}_2^2(\tau)}} = -w \frac{c}{|c|} \frac{\mathbf{x}_1^*(\tau_f)}{R} \\ \mathbf{u}_2^*(\tau) &= -w \frac{\boldsymbol{\lambda}_2(\tau)}{\sqrt{\boldsymbol{\lambda}_1^2(\tau) + \boldsymbol{\lambda}_2^2(\tau)}} = -w \frac{c}{|c|} \frac{\mathbf{x}_2^*(\tau_f)}{R} \\ \mathbf{v}^*(\tau) &= \operatorname{sgn} \boldsymbol{\lambda}_1(\tau) = \operatorname{sgn} c\mathbf{x}_1^*(\tau_f) \end{aligned} \tag{3.87}$$

where $\mathbf{u}_1^*(\tau) \triangleq p_1^*(\mathbf{x}^*(\tau))$, $\mathbf{u}_2^*(\tau) \triangleq p_2^*(\mathbf{x}^*(\tau))$, $\mathbf{v}^*(\tau) \triangleq e^*(\mathbf{x}^*(\tau))$, and so, by condition (b) of Theorem 3.2,

$$1 - w\sqrt{\boldsymbol{\lambda}_1^2(\tau) + \boldsymbol{\lambda}_2^2(\tau)} + |\boldsymbol{\lambda}_1(\tau)| = 0 \tag{3.88}$$

From (3.85), (3.86) and (3.88) we deduce

$$\frac{|\boldsymbol{\lambda}_1(\tau_f)|}{\sqrt{\boldsymbol{\lambda}_1^2(\tau_f) + \boldsymbol{\lambda}_2^2(\tau_f)}} = \frac{|\mathbf{x}_1^*(\tau_f)|}{R} < w \tag{3.89}$$

Hence,

$$|x_1| < Rw \tag{3.90}$$

must be satisfied on the *useable part of the target*, that is on the subset of the target which can be reached by a path π^*. The projection on the x_1–x_2 plane of the useable part of θ is shown on Fig. 3.9. It is $P'Q' \cup P''Q''$, where

$$P'Q' \triangleq \{(x_1, x_2)\colon\ \ x_1^2 + x_2^2 = R^2, \ \ |x_1| < Rw, \ \ x_2 > 0\}$$

$$P''Q'' \triangleq \{(x_1, x_2)\colon\ \ x_1^2 + x_2^2 = R^2, \ \ |x_1| < Rw, \ \ x_2 < 0\}$$

Along a path π^* that reaches θ at point $x^f = \mathbf{x}^*(\tau_f)$ Eqs. (3.78) may be rewritten as

$$\begin{aligned}
\frac{dx_1}{d\tau} &= -w\frac{c}{|c|}\frac{\mathbf{x}_1^*(\tau_f)}{R} + \operatorname{sgn} c\mathbf{x}_1^*(\tau_f) \\
\frac{dx_2}{d\tau} &= -w\frac{c}{|c|}\frac{\mathbf{x}_2^*(\tau_f)}{R} \\
\frac{dx_3}{d\tau} &= 1
\end{aligned} \tag{3.91}$$

Upon integration of (3.91) we find that a path π^* that reaches θ at point $x^f = \mathbf{x}^*(\tau_f)$ is a straight line whose projection on the x_1–x_2 plane is given by

$$x_2 = ax_1 + b \tag{3.92}$$

where

$$a \triangleq \frac{\mathbf{x}_2^*(\tau_f)}{\mathbf{x}_1^*(\tau_f) - \dfrac{R}{w}\dfrac{|c|}{c}\operatorname{sgn} c\mathbf{x}_1^*(\tau_f)}$$

$$b \triangleq -\frac{R}{w}\frac{|c|}{c}\, a \operatorname{sgn} c\mathbf{x}_1^*(\tau_f)$$

A straight line given by Eq. (3.92) in the x_1–x_2 plane intersects the x_1-axis at point P: $x_1 = R/w$, or at point Q: $x_1 = -R/w$, depending on the sign of $\mathbf{x}_1^*(\tau_f)$. One can readily verify that P and Q belong to the tangents to the circle given by $x_1^2 + x_2^2 = R^2$ at the limit points P', P'' and Q', Q'', respectively, of its useable part,† as shown on Fig. 3.9.

Furthermore, from the second of Eqs. (3.91) it follows that a path π^* can reach θ only if $c > 0$. Of course, a point of the useable part of θ for which $x_2 > 0$, can be reached by a path π^* only if $dx_2/d\tau < 0$; and a point for which $x_2 < 0$ can be reached only if $dx_2/d\tau > 0$. This is clear on Fig. 3.9 where the projections on the x_1–x_2 plane of the useable part of θ and of a family of paths are shown. This condition implies the choice of the positive root in (3.85); namely, $c/|c| = 1$.

In summary, given any point of the useable part of θ, a path π^* that reaches θ at that point is obtained by integrating Eqs. (3.91) backward. Its projection on the x_1–x_2 plane is given by (3.92), with $c/|c| = 1$. However, we must stop the integration of Eq. (3.91) at the point where the solution reaches the x_2–x_3 plane. This can be explained as follows: Through a given initial point $x^i = (x_1^i, x_2^i, x_3^i)$ that belongs to both $X^* \cap \text{comp}\,\theta$ and the x_2–x_3 plane, there pass two paths, one for $\boldsymbol{\lambda}_1(\tau) > 0$ and the other for $\boldsymbol{\lambda}_1(\tau) < 0$. They correspond to the same value of τ_f, as is seen readily by integrating the second of Eqs. (3.91); namely,

$$\mathbf{x}_2^*(\tau_f)[1 + (w/R)\tau_f] = x_2^i \tag{3.93}$$

and since $\mathbf{x}_2^*(\tau_f)$ is the same for both paths, τ_f is the same.

If, as we have said before, strategy pair (p^*, e^*) is optimal on X^*, then there exists an optimal path π^* that emanates from each point $x^i \in X^*$, and the value of τ_f computed along this path is the Value of the game for play starting at x^i. Now, if x^i belongs to both $X^* \cap \text{comp}\,\theta$ and the x_2–x_3 plane, then for the same optimal strategy pair there exist two optimal paths emanating from x^i. More precisely, there is no uniqueness of the solution of the state equations

$$\frac{dx_1}{d\tau} = p_1^*(x) + e^*(x), \qquad \frac{dx_2}{d\tau} = p_2^*(x), \qquad \frac{dx_3}{d\tau} = 1$$

in an open ball with center x^i, no matter how small this ball is. But that implies that there does not exist a neighborhood $\Delta(x^i)$ of x^i on which p^* and e^* are of class C^1. Hence, the portion of the x_2–x_3 plane that belongs to $X^* \cap \text{comp}\,\theta$ is a discontinuity manifold for the strategy pair (p^*, e^*).

† Here, we mean by "useable part" the projection of the useable part of θ on the x_1–x_2 plane.

In view of the uniqueness of

$$\boldsymbol{\lambda}_1(\tau) = \frac{\partial V^*(\mathbf{x}^*(\tau))}{\partial x_1}, \qquad \boldsymbol{\lambda}_2(\tau) = \frac{\partial V^*(\mathbf{x}^*(\tau))}{\partial x_2} \qquad \text{and} \qquad \boldsymbol{\lambda}_3(\tau) = \frac{\partial V^*(\mathbf{x}^*(\tau))}{\partial x_3}$$

at points $\mathbf{x}^*(\tau)$, $\tau \neq \tau_f$, which do not belong to a discontinuity manifold, and of the continuity of $\boldsymbol{\lambda}$ along non-null paths π^* which do not intersect a discontinuity manifold, we can conclude that paths with $\boldsymbol{\lambda}_1(\tau) > 0$ emanate from initial states with $x_1^0 > 0$, while those with $\boldsymbol{\lambda}_1(\tau) < 0$ issue from initial states with $x_1^0 < 0$.

Finally, we see that the projection of X^* on the x_1–x_2 plane is the union of regions $Q'P'S'$, $Q''P''S''$ and the projection of target θ, as shown on Fig. 3.9.

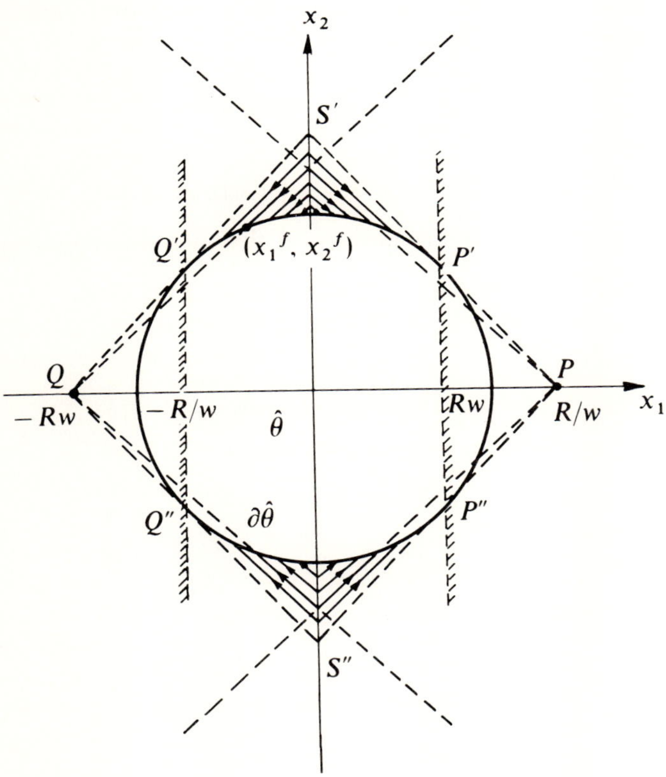

FIG. 3.9. Example 3.1. Projection on the x_1–x_2 plane of the target, useable part of the target, discontinuity manifold and paths [$\hat{\theta}$: projection on the x_1–x_2 plane of θ].

3.20 EXAMPLE 3.2

Next let us consider a game described by

$$\frac{dx_1}{d\tau} = 0, \qquad \frac{dx_2}{d\tau} = u + v, \qquad \frac{dx_3}{d\tau} = 1 \tag{3.94}$$

with constraints

$$\begin{aligned} K_u: &\quad -1 \leqslant u \leqslant 0 \\ K_v: &\quad -1 \leqslant v \leqslant 0 \end{aligned} \tag{3.95}$$

Target θ is defined by

$$\theta \triangleq \{x = (x_1, x_2, x_3): \quad x_2 \leqslant 0\} \tag{3.96}$$

The performance index associated with a path π^{is}, represented by $\mathbf{x}: \quad \tau \to x = \mathbf{x}(\tau)$, $\tau \in [0, \tau_s]$, is

$$\int_0^{\tau_s} (1 - \mathbf{x}_1(\tau)\mathbf{x}_2(\tau))\, d\tau \tag{3.97}$$

The $\mathscr{H}$-function is

$$\mathscr{H}(\lambda, x, u, v) = 1 - x_1 x_2 + \lambda_2(u + v) + \lambda_3 \tag{3.98}$$

Since the constraints are independent of the state, the adjoint equations are

$$\frac{d\lambda_1}{d\tau} = x_2, \qquad \frac{d\lambda_2}{d\tau} = x_1, \qquad \frac{d\lambda_3}{d\tau} = 0 \tag{3.99}$$

so that

$$\boldsymbol{\lambda}_3(\tau) \equiv \text{constant} \tag{3.100}$$

From the transversality condition we obtain

$$\boldsymbol{\lambda}_1(\tau_f) = \boldsymbol{\lambda}_3(\tau_f) = 0 \tag{3.101}$$

so that, in view of (3.100),

$$\boldsymbol{\lambda}_3(\tau) \equiv 0 \tag{3.102}$$

From condition (a) of Theorem 3.2 we obtain

$$\begin{aligned} \mathbf{u}^*(\tau) &= -1 &\quad \text{if} \quad &\boldsymbol{\lambda}_2(\tau) > 0 \\ \mathbf{u}^*(\tau) &= 0 &\quad \text{if} \quad &\boldsymbol{\lambda}_2(\tau) < 0 \\ \mathbf{v}^*(\tau) &= 0 &\quad \text{if} \quad &\boldsymbol{\lambda}_2(\tau) > 0 \\ \mathbf{v}^*(\tau) &= -1 &\quad \text{if} \quad &\boldsymbol{\lambda}_2(\tau) < 0. \end{aligned} \tag{3.103}$$

and so, by condition (b),

$$1 - \mathbf{x}_1^*(\tau)\mathbf{x}_2^*(\tau) - \boldsymbol{\lambda}_2(\tau) = 0 \tag{3.104}$$

Along a path π^* that reaches θ at point $x^f = \mathbf{x}^*(\tau_f)$ Eqs. (3.94) may be rewritten as

$$\frac{dx_1}{d\tau} = 0, \qquad \frac{dx_2}{d\tau} = -1, \qquad \frac{dx_3}{d\tau} = 1 \tag{3.105}$$

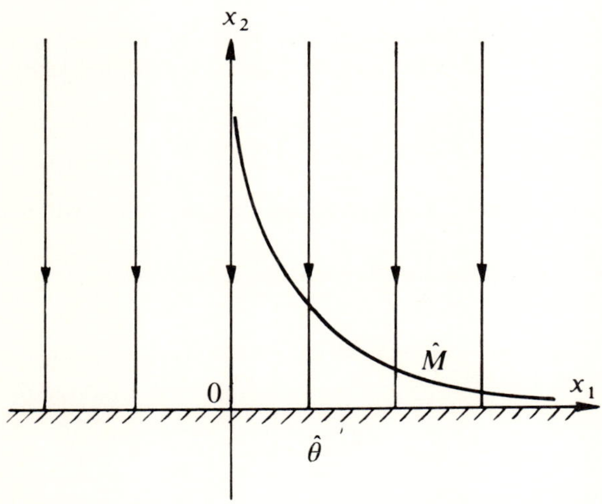

FIG. 3.10. Example 3.2. Projection on the x_1–x_2 plane of target θ, paths and discontinuity manifold. [$\hat{M}$: projection on the x_1–x_2 plane of M, $\hat{\theta}$: projection on the x_1–x_2 plane of θ.]

Upon integration of (3.105) we find that a path π^* that emanates from $x^0 = (x_1^0, x_2^0, x_3^0)$, $x_2^0 > 0$, and that reaches θ at point $x^f = \mathbf{x}^*(\tau_f)$, is a straight line whose projection on the x_1–x_2 plane is given by

$$\begin{aligned} \mathbf{x}_1^*(\tau) &= x_1^0 \\ \mathbf{x}_2^*(\tau) &= \tau^f - \tau \end{aligned} \tag{3.106}$$

By (3.104) the sign of $\boldsymbol{\lambda}_2(\tau)$ switches on the surface

$$M \triangleq \{x = (x_1, x_2, x_3):\ \ 1 - x_1 x_2 = 0\} \tag{3.107}$$

Consequently, the strategy pair that satisfies the necessary conditions for optimality is given by

$$\left.\begin{aligned} p^*(x) &= 0 \\ e^*(x) &= -1 \end{aligned}\right\} \quad \text{for} \quad 1 - x_1x_2 < 0$$
$$\left.\begin{aligned} p^*(x) &= -1 \\ e^*(x) &= 0 \end{aligned}\right\} \quad \text{for} \quad 1 - x_1x_2 > 0 \tag{3.108}$$

Thus, M is a discontinuity manifold. Paths π^* emanating from points $x^0 = (x_1{}^0, x_2{}^0, x_3{}^0)$, $x_1{}^0 > 0$, $x_2{}^0 > 0$, cross M. The jump condition (3.68) becomes

$$\boldsymbol{\lambda}(\tau_c - 0) = \boldsymbol{\lambda}(\tau_c + 0) \tag{3.109}$$

since $c_3/c_1 = 0$. Hence, $\boldsymbol{\lambda}$ is continuous at the point where a path π^* crosses M.

The Value of the game for play starting at point x is

$$V^*(x) = \int_0^{\tau_f} (1 - \mathbf{x}_1{}^*(\tau)\mathbf{x}_2{}^*(\tau))\, d\tau = x_2 - \tfrac{1}{2}x_1x_2{}^2 \tag{3.110}$$

Hence,

$$\Phi(\mathbf{x}) \triangleq x_0 + V^*(x) = x_0 + x_2 - \tfrac{1}{2}x_1x_2{}^2$$

Note that $\Phi(\mathbf{x})$ is independent of the x_3-coordinate. A surface $\Phi(\mathbf{x}) = C$ is shown on Fig. 3.11 in x_0–x_1–x_2 space.

Finally,

$$\begin{aligned} \boldsymbol{\lambda}_1(\tau) &= \frac{\partial V^*(\mathbf{x}^*(\tau))}{\partial x_1} = -\tfrac{1}{2}(\mathbf{x}_2{}^*(\tau))^2 \\ \boldsymbol{\lambda}_2(\tau) &= \frac{\partial V^*(\mathbf{x}^*(\tau))}{\partial x_2} = 1 - \mathbf{x}_1{}^*(\tau)\mathbf{x}_2{}^*(\tau) \\ \boldsymbol{\lambda}_3(\tau) &= \frac{\partial V^*(\mathbf{x}^*(\tau))}{\partial x_3} = 0 \end{aligned} \tag{3.111}$$

These expressions may also be obtained by integrating Eqs. (3.99).

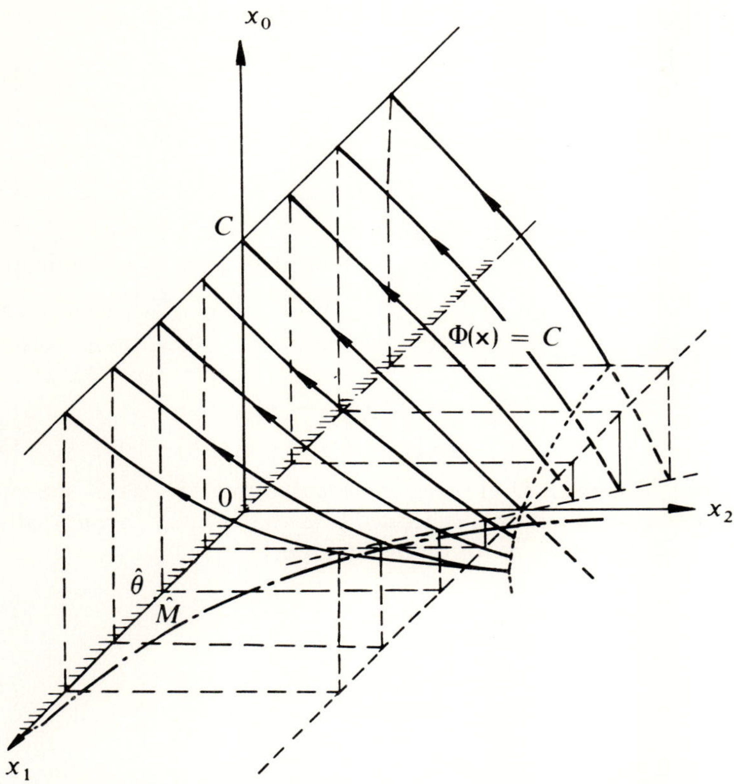

FIG. 3.11. Example 3.2. Projection on the x_0–x_1–x_2 space of game surface $\Sigma(C)$. [$\hat{M}$: projection on the x_1–x_2 plane of M; $\hat{\theta}$: projection on the x_1–x_2 plane of θ.]

3.21 EXAMPLE 3.3

Next let us consider a game described by

$$\frac{dx_1}{d\tau} = u_1\sqrt{x_2} + \frac{w}{2}(v + 1)$$
$$\frac{dx_2}{d\tau} = u_2\sqrt{x_2} + \frac{w}{2}(v - 1) \tag{3.112}$$
$$\frac{dx_3}{d\tau} = 1$$

where w is a positive constant.

The constraints are

$$K_u: \quad u_1^2 + u_2^2 \leqslant 1$$
$$K_v: \quad |v| \leqslant 1 \tag{3.113}$$

Target θ is the closed half-space

$$\theta \triangleq \{x = (x_1, x_2, x_3): \quad x_1 \leqslant 0\} \tag{3.114}$$

The payoff is time to termination.

In this problem $G = E^3$. Let us note that there is no optimal path that emanates from an initial state in the domain $\mathscr{D} \triangleq \{x = (x_1, x_2, x_3): x_2 < 0, x_1 > 0\}$.

Here, the $\mathscr{H}$-function is

$$\mathscr{H}(\lambda, x, u, v) = 1 + \sqrt{x_2}(\lambda_1 u_1 + \lambda_2 u_2) + (\tfrac{1}{2}w)(\lambda_1(v+1) + \lambda_2(v-1)) + \lambda_3 \tag{3.115}$$

Since constraints (3.113) are independent of the state, the adjoint equations along a path π^*, generated by a strategy pair (p^*, e^*) satisfying the necessary conditions for optimality, are

$$\frac{d\lambda_1}{d\tau} = 0$$
$$\frac{d\lambda_2}{d\tau} = -\frac{u_1}{2\sqrt{x_2}}\lambda_1 - \frac{u_2}{2\sqrt{x_2}}\lambda_2 \tag{3.116}$$
$$\frac{d\lambda_3}{d\tau} = 0$$

From the transversality conditions we have

$$\boldsymbol{\lambda}_2(\tau_f) = \boldsymbol{\lambda}_3(\tau_f) = 0 \tag{3.117}$$

so that, by (3.116),

$$\boldsymbol{\lambda}_3(\tau) \equiv 0 \tag{3.118}$$

From condition (a) of Theorem 3.2 we obtain

$$\mathbf{u}_1^*(\tau) = -\frac{\boldsymbol{\lambda}_1(\tau)}{\sqrt{\boldsymbol{\lambda}_1^2(\tau) + \boldsymbol{\lambda}_2^2(\tau)}}$$
$$\mathbf{u}_2^*(\tau) = -\frac{\boldsymbol{\lambda}_2(\tau)}{\sqrt{\boldsymbol{\lambda}_1^2(\tau) + \boldsymbol{\lambda}_2^2(\tau)}} \tag{3.119}$$
$$\mathbf{v}^*(\tau) = \operatorname{sgn}(\boldsymbol{\lambda}_1(\tau) + \boldsymbol{\lambda}_2(\tau))$$

and, by condition (b),

$$1 - \sqrt{\mathbf{x}_2^*(\tau)}\sqrt{\boldsymbol{\lambda}_1^2(\tau) + \boldsymbol{\lambda}_2^2(\tau)} + \frac{w}{2}|\boldsymbol{\lambda}_1(\tau) + \boldsymbol{\lambda}_2(\tau)| + \frac{w}{2}(\boldsymbol{\lambda}_1(\tau) - \boldsymbol{\lambda}_2(\tau)) = 0 \tag{3.120}$$

Now let us consider two cases:

(i) $\boldsymbol{\lambda}_1(\tau) + \boldsymbol{\lambda}_2(\tau) > 0$; then Eqs. (3.112) may be written

$$\begin{aligned}\frac{dx_1}{d\tau} &= -\sqrt{x_2}\frac{\boldsymbol{\lambda}_1(\tau)}{\sqrt{\boldsymbol{\lambda}_1^2(\tau) + \boldsymbol{\lambda}_2^2(\tau)}} + w\\ \frac{dx_2}{d\tau} &= -\sqrt{x_2}\frac{\boldsymbol{\lambda}_2(\tau)}{\sqrt{\boldsymbol{\lambda}_1^2(\tau) + \boldsymbol{\lambda}_2^2(\tau)}}\\ \frac{dx_3}{d\tau} &= 1\end{aligned} \tag{3.121}$$

and

(ii) $\boldsymbol{\lambda}_1(\tau) + \boldsymbol{\lambda}_2(\tau) < 0$; then Eqs. (3.112) may be written

$$\begin{aligned}\frac{dx_1}{d\tau} &= -\sqrt{x_2}\frac{\boldsymbol{\lambda}_1(\tau)}{\sqrt{\boldsymbol{\lambda}_1^2(\tau) + \boldsymbol{\lambda}_2^2(\tau)}}\\ \frac{dx_2}{d\tau} &= -\sqrt{x_2}\frac{\boldsymbol{\lambda}_2(\tau)}{\sqrt{\boldsymbol{\lambda}_1^2(\tau) + \boldsymbol{\lambda}_2^2(\tau)}} - w\\ \frac{dx_3}{d\tau} &= 1\end{aligned} \tag{3.122}$$

For Case (ii), it follows from (3.117) and (3.120) at $\tau = \tau_f$ that

$$\boldsymbol{\lambda}_1(\tau_f) < 0$$

and

$$\frac{d\mathbf{x}_1^*(\tau_f)}{d\tau} = -\sqrt{\mathbf{x}_2^*(\tau_f)}\frac{\boldsymbol{\lambda}_1(\tau_f)}{|\boldsymbol{\lambda}_1(\tau_f)|} = \frac{1}{|\boldsymbol{\lambda}_1(\tau_f)|}$$

which is in contradiction with the fact that non-null paths which reach θ belong to the half-space $x_1 \geqslant 0$. Hence,

$$\boldsymbol{\lambda}_1(\tau_f) > 0 \tag{3.123}$$

Thus Case (ii) cannot apply at $\tau = \tau_f$.

Condition (3.123) has two consequences. First, it follows from (3.117), (3.120), and (3.123) that

$$(\sqrt{\mathbf{x}_2^*(\tau_f)} - w)\boldsymbol{\lambda}_1(\tau_f) > 0$$

Hence, the subset of the target that can be reached by a path π^*, that is the useable part of the target, must satisfy the condition

$$x_2 > w^2 \tag{3.124}$$

Secondly, if there exist a strategy pair (p^*, e^*) that satisfies the necessary conditions for optimality and a path π^* that reaches θ at point $\mathbf{x}^*(\tau_f)$, then, in view of the continuity of $\boldsymbol{\lambda}$ along every non-null portion of π^* that does not intersect a discontinuity manifold, there exists an interval of time $(\tau_f - \tau_c, \tau_f]$, $\tau_c > 0$, on which $\boldsymbol{\lambda}_1(\tau)$ and $\boldsymbol{\lambda}_1(\tau) + \boldsymbol{\lambda}_2(\tau)$ are strictly positive.

So, let us consider Case (i) first; that is, let $\tau \in (\tau_f - \tau_c, \tau_f]$, and accordingly $\mathbf{v}(\tau) = 1$ in Eq. (3.119). Equation (3.120) becomes

$$1 - \sqrt{\mathbf{x}_2^*(\tau)}\sqrt{\boldsymbol{\lambda}_1^2(\tau) + \boldsymbol{\lambda}_2^2(\tau)} + w\boldsymbol{\lambda}_1(\tau) = 0 \tag{3.125}$$

For $\tau = \tau_f$, we deduce from (3.125) that

$$\boldsymbol{\lambda}_1(\tau_f) = [\sqrt{\mathbf{x}_2^*(\tau_f)} - w]^{-1} \tag{3.126}$$

and so by the first of Eqs. (3.116),

$$\boldsymbol{\lambda}_1(\tau) \equiv \text{constant} = [\sqrt{\mathbf{x}_2^*(\tau_f)} - w]^{-1} \tag{3.127}$$

By substituting expression (3.127) for $\boldsymbol{\lambda}_1(\tau)$ in Eq. (3.125) one readily obtains

$$\boldsymbol{\lambda}_2^2(\tau) = \frac{\mathbf{x}_2^*(\tau_f)/\mathbf{x}_2^*(\tau) - 1}{(\sqrt{\mathbf{x}_2^*(\tau_f)} - w)^2} \tag{3.128}$$

Equation (3.128) requires that

$$\frac{\mathbf{x}_2^*(\tau_f)}{\mathbf{x}_2^*(\tau)} \geqslant 1 \qquad \text{for } \tau \in (\tau_f - \tau_c, \tau_f] \tag{3.129}$$

which implies the choice of the negative root in (3.128). For, if we choose the positive root, then $dx_2/d\tau < 0$ in the second of Eqs. (3.121), which implies that $\mathbf{x}_2^*(\tau)$ is a decreasing function of τ; this, in turn, implies that

$$\mathbf{x}_2^*(\tau) > \mathbf{x}_2^*(\tau_f) \qquad \text{for} \qquad \tau < \tau_f.$$

But that contradicts (3.129). Hence we have

$$\boldsymbol{\lambda}_2(\tau) = -\frac{\sqrt{\mathbf{x}_2^*(\tau_f)/\mathbf{x}_2^*(\tau) - 1}}{\sqrt{\mathbf{x}_2^*(\tau_f)} - w} \qquad \tau \in (\tau_f - \tau_c, \tau_f] \tag{3.130}$$

By substituting (3.127) and (3.130) in Eqs. (3.121) we obtain

$$\begin{aligned} \frac{dx_1}{d\tau} &= -\frac{x_2}{\sqrt{\mathbf{x}_2^*(\tau_f)}} + w \\ \frac{dx_2}{d\tau} &= \sqrt{x_2}\sqrt{1 - \frac{x_2}{\mathbf{x}_2^*(\tau_f)}} \\ \frac{dx_3}{d\tau} &= 1 \end{aligned} \tag{3.131}$$

Next, let $\tau' \triangleq \tau_f - \tau$. By integrating Eq. (3.131) backward one obtains

$$\begin{aligned} x_1 &= \frac{\tau'\sqrt{\mathbf{x}_2^*(\tau_f)}}{2} + \frac{\mathbf{x}_2^*(\tau_f)}{2}\sin\frac{\tau'}{\sqrt{\mathbf{x}_2^*(\tau_f)}} - w\tau' \\ x_2 &= \frac{\mathbf{x}_2^*(\tau_f)}{2}\left(1 + \cos\frac{\tau'}{\sqrt{\mathbf{x}_2^*(\tau_f)}}\right) \end{aligned} \tag{3.132}$$

provided that $\tau' \leqslant \pi\sqrt{\mathbf{x}_2^*(\tau_f)}$.

From (3.127) and (3.130) we have

$$\boldsymbol{\lambda}_1(\tau) + \boldsymbol{\lambda}_2(\tau) = \frac{1 - \sqrt{\dfrac{\mathbf{x}_2^*(\tau_f)}{\mathbf{x}_2^*(\tau)} - 1}}{\sqrt{\mathbf{x}_2^*(\tau_f)} - w}$$

Hence, the sign of $\boldsymbol{\lambda}_1(\tau) + \boldsymbol{\lambda}_2(\tau)$ becomes negative (in the backward integration of Eqs. 3.131) when $\mathbf{x}_2^*(\tau) = \mathbf{x}_2^*(\tau_f)/2$, that is for $\tau' \triangleq \tau_f - \tau = \sqrt{\mathbf{x}_2^*(\tau_f)}\pi/2$. By substituting in Eqs. (3.132) one obtains the parametric equations of a parabola $\hat{M}$; namely,

$$x_1 = \left(\frac{\pi}{4} + \frac{1}{2}\right)s - w\frac{\pi}{2}\sqrt{s}, \quad x_2 = \frac{s}{2}$$

where

$$s \triangleq \mathbf{x}_2^*(\tau_f)$$

By eliminating s between these equations we see that the sign of $\boldsymbol{\lambda}_1(\tau) + \boldsymbol{\lambda}_2(\tau)$ becomes negative on the surface

$$M \triangleq \left\{x = (x_1, x_2, x_3):\quad x_1 - \left(\frac{\pi}{2} + 1\right)x_2 + w\frac{\pi}{\sqrt{2}}\sqrt{x_2} = 0\right\} \tag{3.133}$$

In the region where $\boldsymbol{\lambda}_1(\tau) + \boldsymbol{\lambda}_2(\tau) < 0$ the control vector $\mathbf{v}^*(\tau)$, given by the third of Eqs. (3.119), is $\mathbf{v}^*(\tau) = -1$.

The trajectory equations can be deduced from Eqs. (3.122) by arguments similar to the ones employed in the case treated above. We shall let the reader verify that the paths thus obtained, by piecing together the part of the solution corresponding to $\mathbf{v}^*(\tau) = +1$ and the one corresponding to $\mathbf{v}^*(\tau) = -1$, cross M.

Let $x^c = \mathbf{x}^*(\tau_c)$ be a point where a path π^* crosses M. Since $\mathbf{v}^*(\tau)$ has a discontinuity at $\tau = \tau_c$, $\mathbf{v}^*(\tau) = e^*(\mathbf{x}^*(\tau))$, and $\mathbf{x}^*$ is a continuous function of τ on $[0, \tau_f]$, there does not exist a neighborhood $\Delta(x^c)$ of x^c on which $e^*(x)$ is of class C^1. Otherwise $\mathbf{v}^*$ would be continuous at $\tau = \tau_c$. Hence, M is a discontinuity manifold. The projection $\hat{M}$ of M on the x_1–x_2 plane is shown on Fig. 3.12.

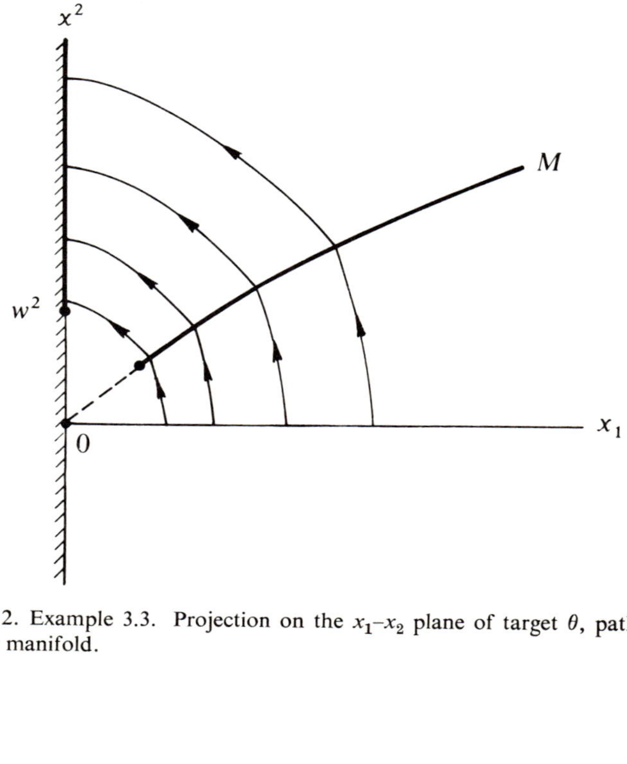

FIG. 3.12. Example 3.3. Projection on the x_1–x_2 plane of target θ, paths and discontinuity manifold.

3.22 SUFFICIENCY CONDITIONS

In this section we shall deduce a sufficiency theorem for optimal strategies. Before doing so, let us introduce some definitions.

Suppose that $D \triangleq \{X_1, X_2, \ldots, X_K\}$ constitutes a decomposition of X, and consider the sets

$$M \triangleq X - \bigcup_{\sigma=1}^{K} X_\sigma \tag{3.134}$$

and

$$M_k \triangleq \bar{X}_\alpha \cap \bar{X}_\beta \cap \cdots \cap \bar{X}_\delta \cap X \tag{3.135}$$

where $X_\alpha, X_\beta, \ldots, X_\delta \in D$.

It is quite evident that M_k can be constructed so that

$$M_k \subseteq M \tag{3.136}$$

and, in fact,

$$M = \bigcup_{k=1}^{L<\infty} M_k \tag{3.137}$$

Definition 3.9. Given a decomposition D of X, we shall say that strategy pair (p, e) is *well behaved* at $x^i \in X$ relative to D, if (p, e) generates a terminating path π^{if} represented by $\mathbf{x}\colon \tau \to x = \mathbf{x}(\tau)$, $\tau \in [0, \tau_f]$, $\mathbf{x}(0) = x^i$, such that

(i) $\mathbf{x}(\tau) \in X$ for all $\tau \in [0, \tau_f)$;

(ii) for all $\tau' \in [0, \tau_f]$ there exists an interval $[\tau_a, \tau_b]$ such that $\tau' \in [\tau_a, \tau_b]$ and $\mathbf{x}(\tau) \in X_\sigma \in D$ for all $\tau \in (\tau_a, \tau_b)$ or $\mathbf{x}(\tau) \in M_k \subseteq M$ for all $\tau \in (\tau_a, \tau_b)$;

(iii) Assumption 3.3 if fulfilled.

Now we are ready to prove

Theorem 3.3. *Strategy pair (p^*, e^*), with corresponding solution $\mathbf{x}^*\colon \tau \to x = \mathbf{x}^*(\tau)$, $\tau \in [0, \tau_f^*]$, $\mathbf{x}^*(0) = x^i$, is optimal with respect to all well-behaved (at $x^i \in X$ relative to the decomposition D below) strategy pairs (p, e), with corresponding solutions $\mathbf{x}\colon \tau \to x = \mathbf{x}(\tau)$, $\tau \in [0, \tau_f]$, $\mathbf{x}(0) = x^i$, if it is playable at all $x^i \in X$ and if there exist a decomposition*

D of X and a function $V^*\colon \quad x \to V^*(x)$ *which*

(i) *is continuous on* X*;*

(ii) *is of class* C^1 *on each* $X_\sigma \in D$*;*

(iii) *on every non-empty* $M_k \subseteq M$ *agrees with a function—say* $V_k{}^*$*— that is of class* C^1 *on a domain containing* M_k*; and*

(iv) $V^*(x) = 0$ *for all* $x \in \theta$*;*

such that

(a)† $$\int_0^{\tau_f{}^*} [f_0(\mathbf{x}^*(\tau), p^*(\mathbf{x}^*(\tau)), e^*(\mathbf{x}^*(\tau))) + \operatorname{grad} V^*(\mathbf{x}^*(\tau)) \cdot f(\mathbf{x}^*(\tau), p^*(\mathbf{x}^*(\tau)), e^*(\mathbf{x}^*(\tau)))]\, d\tau = 0,$$
$$\forall x^i \in X;$$

(b) $$f_0(x, p^*(x), v) + \operatorname{grad} V^*(x) \cdot f(x, p^*(x), v) \leqslant 0, \qquad \forall x \in X - \theta, \qquad \forall v \in K_v(x);$$

(c) $$f_0(x, u, e^*(x)) + \operatorname{grad} V^*(x) \cdot f(x, u, e^*(x)) \geqslant 0, \qquad \forall x \in X - \theta, \qquad \forall u \in K_u(x);$$

where $V^*(x) = V_k{}^*(x)$ *if* $x \in M_k$.

Consider a path π^{if} that is generated by a well-behaved strategy pair (p, e). Since (p, e) is well-behaved, interval $[0, \tau_f]$ contains a finite or at most a countable number of open intervals $(0, \tau_1)$, (τ_1, τ_2), . . . , $(\tau_{N-1}, \tau_N = \tau_f)$ such that $\mathbf{x}(\tau)$ belongs either to an $X_\sigma \in D$ or to an $M_k \subseteq M$ during any one such open interval.

Thus, integral

$$I(x^i) \triangleq \int_{\pi^{if}} \operatorname{grad} V^*(x) \cdot dx \tag{3.138}$$

is composed of a finite or at most countable number of integrals; namely, by (i) and (ii) of the theorem,

$$I(x^i) = \int_{\pi^{i1}} \operatorname{grad} V^*(x) \cdot dx + \int_{\pi^{12}} \operatorname{grad} V^*(x) \cdot dx + \cdots + \int_{\pi^{N-1,f}} \operatorname{grad} V^*(x) \cdot dx \tag{3.139}$$

† This condition is implied by (b) and (c).

where

$$\mathbf{x}(\tau) \in \pi^{i1} \subset \pi^{if} \quad \text{for} \quad \tau \in (0, \tau_1)$$
$$\mathbf{x}(\tau) \in \pi^{12} \subset \pi^{if} \quad \text{for} \quad \tau \in (\tau_1, \tau_2)$$
$$\vdots$$
$$\mathbf{x}(\tau) \in \pi^{N-1,f} \subset \pi^{if} \quad \text{for} \quad \tau \in (\tau_{N-1}, \tau_f)$$

and where, according to (iii) of the theorem, $V^*(x) = V_k^*(x)$ if $x \in M_k$.

Then it follows from (i) and (iv) of the theorem that

$$I(x^i) = -V^*(x^i) \tag{3.140}$$

Consequently, integral $I(x^i)$ is invariant with the choice of (p, e). In view of this conclusion, together with condition (a) of the theorem and state equations (3.1), we have

$$\int_0^{\tau_f^*} f_0(\mathbf{x}^*(\tau), p^*(\mathbf{x}^*(\tau)), e^*(\mathbf{x}^*(\tau)))\, d\tau = V^*(x^i) \tag{3.141}$$

Next let us consider a well-behaved strategy pair (p^*, e) generating a P-optimal path π_P^{if} represented by $\mathbf{x}$: $\tau \to x = \mathbf{x}(\tau)$, $\tau \in [0, \tau_f]$, $\mathbf{x}(0) = x^i$, and form the cost difference

$$\Delta \triangleq \int_0^{\tau_f} f_0(\mathbf{x}(\tau), p^*(\mathbf{x}(\tau)), e(\mathbf{x}(\tau)))\, d\tau - \int_0^{\tau_f^*} f_0(\mathbf{x}^*(\tau), p^*(\mathbf{x}^*(\tau)), e^*(\mathbf{x}^*(\tau)))\, d\tau \tag{3.142}$$

Now add the following null-term to the right-hand side of (3.142):

$$\int_{\pi_P^{if}} \operatorname{grad} V^*(x) \cdot dx + V^*(x^i) \tag{3.143}$$

Since

$$\int_{\pi_P^{if}} \operatorname{grad} V^*(x) \cdot dx = \int_0^{\tau_f} \operatorname{grad} V^*(\mathbf{x}(\tau)) \cdot f(\mathbf{x}(\tau), p^*(\mathbf{x}(\tau)), e(\mathbf{x}(\tau)))\, d\tau \tag{3.144}$$

it follows that

$$\Delta = \int_0^{\tau_f} [f_0(\mathbf{x}(\tau), p^*(\mathbf{x}(\tau)), e(\mathbf{x}(\tau)))$$

$$+ \operatorname{grad} V^*(\mathbf{x}(\tau)) \cdot f(\mathbf{x}(\tau), p^*(\mathbf{x}(\tau)), e(\mathbf{x}(\tau)))] \, d\tau \qquad (3.145)$$

Now, either π_P^{if} is a null path so that $\Delta = 0$; or π_P^{if} is a non-null path so that $\mathbf{x}(\tau) \in X - \theta$ for all $\tau \in [0, \tau_f)$ and, as a consequence of condition (b) of the theorem,

$$\Delta \leqslant 0 \qquad (3.146)$$

Similarly, by considering an E-optimal path and invoking (c) of the theorem, one can show that $\Delta \geqslant 0$ for a well-behaved (p, e^*). This concludes the proof of Theorem 3.3.

It should be noted that Isaacs' "Verification Theorem" (Ref. 16) constitutes a special case of Theorem 3.3.

We shall leave it to the reader to verify that the strategies deduced in Examples 3.1–3.3 satisfy the conditions of Theorem 3.3 and hence are optimal in the sense of the theorem.

IV *Multistage Quantitative Games*

4.1 PROBLEM STATEMENT

In this chapter we shall consider a multistage quantitative game, that is a quantitative game whose state is characterized by $n - 1$ real numbers x_1, $x_2, \ldots, x_{n-1}$, and by the stage k of the process, where $k \in \{0, 1, \ldots, K\}$

It is convenient to think of the state of the game as a point

$$x \triangleq (x_1, x_2, \ldots, x_n) \in E^n$$

where $x_n \in \{0, 1, \ldots, K\}$. For $x = x^k \triangleq (x_1, x_2, \ldots, x_{n-1}, k)$ we say that the state is at the kth stage.

The state changes, that is, x may take on different values in E^n, as x_n varies from 0 to K. The evolution of the state is governed by a set of difference equations; in particular, we shall suppose that the change of x^k to x^{k+1} depends on x^k and some parameters, the control variables $u_1, u_2, \ldots, u_r$ and $v_1, v_2, \ldots, v_s$, which the players J_P and J_E, respectively, choose at each stage of play. Let $u \triangleq (u_1, u_2, \ldots, u_s)$ and $v \triangleq (v_1, v_2, \ldots, v_s)$. We shall assume that $u \in U$ and $v \in V$, where U and V are fixed domains of E^r and E^s, respectively. With each change or transfer of the state we shall associate a real number, the cost.

It is our purpose to discuss some geometric aspects of optimal play, and we shall illustrate some consequences of geometric properties by presenting a derivation of necessary conditions for a restricted class of problems. We shall assume, without loss of generality, that the multistage game which we are discussing now is the discrete version of a differential game which has been quantized, and that the stage k is the time variable. Of course, all the results will hold if k is not the time.

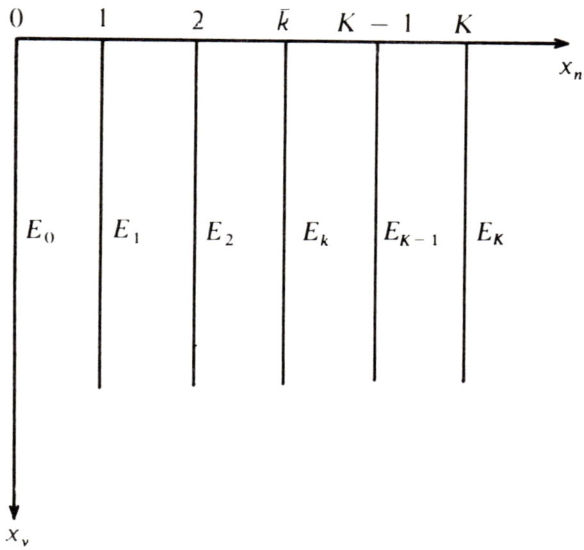

FIG. 4.1. The state space and the sets $E_0, E_1, \ldots, E_K$.

Since x_n takes on integer values on $[0, K]$, we shall consider

$$E_k \triangleq \{x: \quad x_n = k\} \tag{4.1}$$

Let $\mathscr{D}$ be a fixed domain of E^n, such that $G_k \triangleq \mathscr{D} \cap E_k \neq \varnothing$ for $k = 0, 1, \ldots, K$.

In what follows we shall assume that state x belongs to the set $G \triangleq \bigcup_k G_k$, $k \in \{0, 1, \ldots, K\}$.

4.2 STATE EQUATIONS, STRATEGIES AND TARGET θ

We shall consider the state of the game as a function of stage k; namely $\mathbf{x}\colon\ k \to x^k = \mathbf{x}(k)$ on $\{i, i+1, \ldots, j\}$, $0 \leqslant i \leqslant j \leqslant K$, $\mathbf{x}(k) = (\mathbf{x}_1(k), \ldots, \mathbf{x}_{n-1}(k), k)$. If $i \neq j$ we shall require that

$$\mathbf{x}(k+1) - \mathbf{x}(k) = f^k(\mathbf{x}(k), \mathbf{u}(k), \mathbf{v}(k)) \tag{4.2}$$

for $k = i, i+1, \ldots, j-1$, with $\mathbf{u}$ and $\mathbf{v}$ being functions of k,

$$\mathbf{u}\colon\ k \to u = \mathbf{u}(k), \qquad \mathbf{v}\colon\ k \to v = \mathbf{v}(k)$$

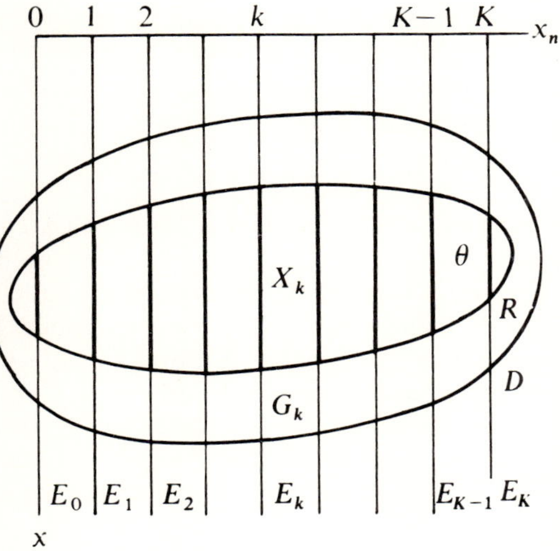

FIG. 4.2. The sets X_k and G_k; $\theta = X_K$.

where k belongs to $\{i, i + 1, \ldots, j - 1\}$, and

$$f^k(x, u, v) \triangleq (f_1^k(x, u, v), f_2^k(x, u, v), \ldots, f_n^k(x, u, v))$$

We shall assume that the functions f_ν^k, $\nu = 1, 2, \ldots, n - 1$, are of class C^1 on $\mathscr{D} \times U \times V$, for $k = 0, 1, \ldots, K - 1$. Of course, since $x_n^k = k$, we have $f_n^k(x, u, v) \equiv 1$.

Let there be a fixed domain $R \subseteq \mathscr{D}$ such that $X_k \triangleq R \cap E_k \neq \varnothing$ for $k = 0, 1, \ldots, K$. Let $X \triangleq \bigcup_k X_k$, $k \in \{0, 1, \ldots, K - 1\}$.

We shall consider strategies to be functions of x, p: $x \to p(x)$ and e: $x \to e(x)$, $x \in X$, belonging to prescribed classes, such that

$$p(x) \in K_u(x) \subseteq U, \qquad e(x) \in K_v(x) \subseteq V$$

where $K_u(x)$ and $K_v(x)$ are given subsets of U and V, respectively, which may depend on x.

We shall be interested in transferring the state of the game from an initial state $\mathbf{x}(i) = x^i$ to any one in prescribed set of states $\theta \triangleq X_K$ by means of a strategy pair (p, e), where $\mathbf{u}(k) = p(\mathbf{x}(k))$ and $\mathbf{v}(k) = e(\mathbf{x}(k))$.

4.3 DESCRIBING CURVES AND PATHS IN STATE SPACE

For given strategy pair (p, e) and an initial state $\mathbf{x}(i) = x^i$ in G, Eq. (4.2) defines a unique sequence of states

$$\Gamma = \{\mathbf{x}(i), \mathbf{x}(i+1), \ldots, \mathbf{x}(i+d)\}$$

where $i + d$ is the largest value of stage k for which $\mathbf{x}(i + d)$ is defined. Of course, $i + d \leqslant K$.

If $\mathbf{x}(i) = x^i$ does not belong to X, or if $i = K$, or both, $\mathbf{x}(i + 1)$ is not defined; that is, $d = 0$ and Γ reduces to the single point x^i.

We shall call function $\mathbf{x}$: $k \to x^k = \mathbf{x}(k) \in \Gamma, k \in \{i, i+1, \ldots, i+d\}$ the *maximal solution* of Eq. (4.2) for given initial state x^i and given strategy pair (p, e). Of course, the value of d is defined by x^i, p and e. For given x^i and strategy pair (p, e), there is a unique maximal solution.

Replacing k by the new parameter $\tau = k - i$, the maximal solution $\mathbf{x}$ becomes

$$\chi\colon \quad \tau \to x^k = \chi(\tau), \qquad \tau \in \{0, 1, \ldots, d\}$$

where $\chi(\tau) = \mathbf{x}(\tau + i)$. Then we let $\gamma(x^i, p, e) \triangleq \chi$. Since $\gamma(x^i, p, e)$ is defined for all $x^i \in G$ and for all strategy pairs (p, e), mapping γ is defined.

The rules of the game consist of G, the sets of strategies and γ.

Clearly, Γ is the describing curve emanating from x^i and generated by (p, e). One can readily verify that Assumption 1.1 is satisfied.

From the definition of a path it follows that any subsequence π^{ij} of Γ, namely

$$\pi^{ij} = \{\mathbf{x}(i), \mathbf{x}(i+1), \ldots, \mathbf{x}(j)\} \qquad j = i + l, \quad l \in \{0, 1, \ldots, d\}$$

is a *path*.

A point $\mathbf{x}(k)$, $k \in \{0, 1, \ldots, K\}$, is a null path π^0.

4.4 COST OF TRANSFER

Next let us consider the transfer of the state from a point $\mathbf{x}(i) = x^i$ to a point $\mathbf{x}(j) = x^j$, along a path π^{ij} generated by a strategy pair (p, e). With each such transfer we shall associate a *cost* $\mathscr{V}(x^i, x^j;\ p, e, \pi^{ij})$.

We shall suppose that

(i) the cost of transfer from x^k to x^{k+1} along a non-null path π^{ij} is given by

$$f_0^k(\mathbf{x}(k), \mathbf{u}(k), \mathbf{v}(k)) \tag{4.3}$$

where

$$\mathbf{u}(k) = p(\mathbf{x}(k)), \qquad \mathbf{v}(k) = e(\mathbf{x}(k))$$

and the functions f_0^k are of class C^1 on $\mathscr{D} \times U \times V$ for $k = 0, 1, \ldots, K - 1$;

(ii) the cost of transfer from x^i to x^j, $j = i + l$, along π^{ij} is given by

$$\mathscr{V}(x^i, x^j; p, e, \pi^{ij}) \triangleq \sum_{k=i}^{i+l-1} f_0^k(\mathbf{x}(k), \mathbf{u}(k), \mathbf{v}(k)) \tag{4.4}$$

(iii) the cost of transfer associated with a null path is zero; that is,

$$\mathscr{V}(x^k, x^k, p, e, \pi^0) = 0 \tag{4.5}$$

for all $k \in \{0, 1, \ldots, K\}$ and for all strategy pairs (p, e).

Hence Assumption 2.1 is satisfied. It is clear that Assumption 2.2 is also satisfied.

4.5 AUGMENTED STATE SPACE AND PATHS IN E^{n+1}

Next we introduce

Assumption 4.1. There exists a strategy pair (p^*, e^*) that is optimal on X.

By (4.5), strategy pair (p^*, e^*) is optimal on $X^* \triangleq X \cup \theta = \bigcup_k X_k$, $k \in \{0, 1, \ldots, K\}$.

Now let us consider points

$$\mathbf{x} = (x_0, x_1, \ldots, x_n) = (x_0, x) \in E^{n+1}$$

where $E^{n+1} \triangleq E^n \times \{x_0\}$ is the *augmented state space*, and let $\mathbf{x}_0$: $k \to x_0 = \mathbf{x}_0(k)$, $k \in \{i, i + 1, \ldots, j\}$, $0 \leqslant i < j \leqslant K$, be a function satisfying difference equation

$$x_0(k + 1) = x_0(k) + f_0^k(x(k), u(k), v(k)) \tag{4.6}$$

for $k = i, i + 1, \ldots, j - 1$.

Equations (4.2) and (4.6) constitute a set of $n + 1$ scalar difference equations which we shall write in vector form

$$\mathbf{x}(k + 1) - \mathbf{x}(k) = \mathsf{f}^k(\mathbf{x}(k), \mathbf{u}(k), \mathbf{v}(k)) \tag{4.7}$$

where $\mathbf{x}$: $k \to \mathbf{x}^k = \mathbf{x}(k)$ on $\{i, i + 1, \ldots, j\}$, $\mathbf{x}^k \triangleq (x_0, x_1, \ldots, x_{n-1}, k)$,

$\mathbf{x} = (\mathbf{x}_0, \mathbf{x}_1, \ldots, \mathbf{x}_{n-1}, \mathbf{x}_n)$, and

$$\mathsf{f}^k(\mathbf{x}, u, v) \triangleq (f_0^k(x, u, v), f_1^k(x, u, v), \ldots, f_{n-1}^k(x, u, v), 1)$$

The definition of paths in E^{n+1} will be the same as that in Chapter 2. Now, from the definition of cost it follows that the motion of state $\mathbf{x}$ along a path in E^{n+1} is governed by Eq. (4.7), and

(i) if points $\mathbf{x}(k)$ and $\mathbf{x}(k+1)$ belong to a path $\Pi_P^{ij}(C)$ we have

$$\mathbf{x}(k+1) - \mathbf{x}(k) = \mathsf{f}^k(\mathbf{x}(k), \mathbf{u}^*(k), \mathbf{v}(k)) \tag{4.8}$$

(ii) if points $\mathbf{x}(k)$ and $\mathbf{x}(k+1)$ belong to a path $\Pi_E^{lm}(C)$ we have

$$\mathbf{x}(k+1) - \mathbf{x}(k) = \mathsf{f}^k(\mathbf{x}(k), \mathbf{u}(k), \mathbf{v}^*(k)) \tag{4.9}$$

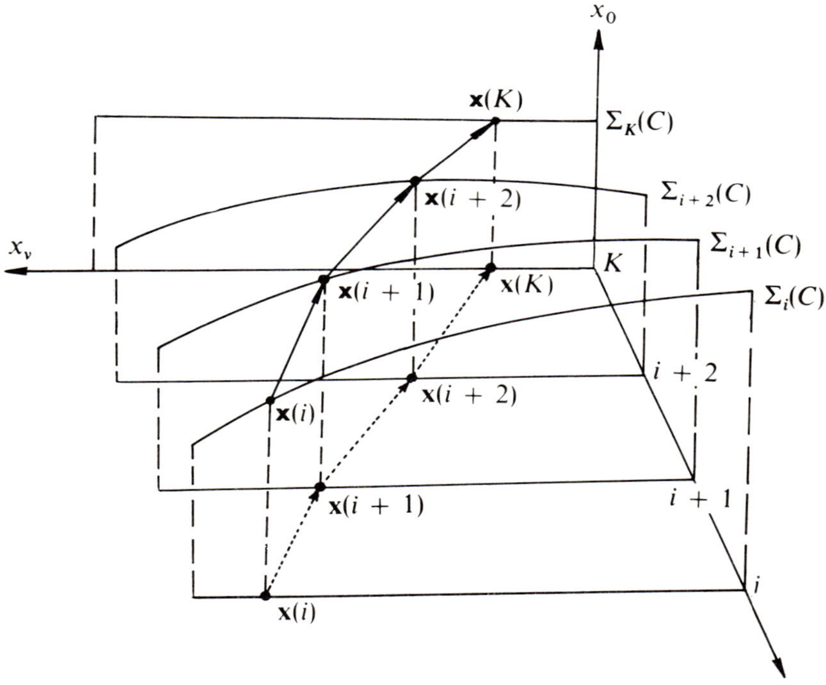

FIG. 4.3. Game surface and path in E^{n+1}

(iii) if points $\mathbf{x}(k)$ and $\mathbf{x}(k+1)$ belong to a path $\Pi_{PE}^{nq}(C)$ we have

$$\mathbf{x}(k+1) - \mathbf{x}(k) = \mathbf{f}^k(\mathbf{x}(k), \mathbf{u}^*(k), \mathbf{v}^*(k)) \tag{4.10}$$

where

$$\mathbf{u}^*(k) = p^*(\mathbf{x}(k)), \qquad \mathbf{v}^*(k) = e^*(\mathbf{x}(k))$$

4.6 SOME PROPERTIES OF A GAME SURFACE

The definition of a game surface will be the same as that in Chapter 2. Furthermore, we shall define a surface $\Sigma_k(C)$, $k \in \{0, 1, \ldots, K\}$, by

Definition 4.1.

$$\Sigma_k(C) \triangleq \Sigma(C) \cap \mathscr{E}_k \tag{4.11}$$

where $\mathscr{E}_k \triangleq E_k \times \{x_0\}$.

A surface $\Sigma_k(C)$, $k \in \{0, 1, \ldots, K\}$, is a single sheeted surface in $\mathscr{X}_k \triangleq X_k \times \{x_0\}$; that is, it is a set of points which are in one-to-one correspondence with the points of X_k.

Definition 4.2.
A surface $\Sigma_k(C)$, $k \in \{0, 1, \ldots, K\}$, separates $\mathscr{X}_k$ into two disjoint sets, namely $A/\Sigma_k(C)$ and $B/\Sigma_k(C)$, defined in

$$A/\Sigma_k(C) \triangleq (A/\Sigma(C)) \cap \mathscr{E}_k$$

$$B/\Sigma_k(C) \triangleq (B/\Sigma(C)) \cap \mathscr{E}_k$$

The equation of $\Sigma_k(C)$ is deduced from (2.5). It is

$$\Phi(\mathsf{x}) \triangleq x_0 + V^*(x) = C, \qquad x \in X_k \tag{4.12}$$

where $V^*(x)$ is the Value of the game for play starting at point x; and, accordingly,

$$\Sigma_k(C) = \{\mathsf{x} = (x_0, x)\colon\ x \in X_k,\ \ \Phi(\mathsf{x}) = C\} \tag{4.13}$$

$$A/\Sigma_k(C) = \{\mathsf{x} = (x_0, x)\colon\ x \in X_k,\ \ \Phi(\mathsf{x}) > C\} \tag{4.14}$$

$$B/\Sigma_k(C) = \{\mathsf{x} = (x_0, x)\colon\ x \in X_k,\ \ \Phi(\mathsf{x}) < C\} \tag{4.15}$$

As the value of parameter C is varied, Eq. (4.12) defines a one-parameter family of surfaces, namely $\{\Sigma_k(C)\}$.

Definition 4.3. For a given C, we shall say that $\mathbf{x}^k \in \Sigma_k(C)$ is an interior point of $\Sigma_k(C)$, if there exists an n-dimensional ball $\Delta(\mathbf{x}^k)$ with center $\mathbf{x}^k$ such that $\Delta(\mathbf{x}^k) \subset \mathscr{X}_k$.

Since R is open in E^n, $X_k \triangleq R \cap E_k$ is open in E_k and $\mathscr{X}_k \triangleq X_k \times \{x_0\}$ is open in $E_k \times \{x_0\}$. Hence, every point of $\Sigma_k(C)$ is an interior point of $\Sigma_k(C)$, $k = 0, 1, \ldots, K$.

4.7 SETS Ω_P AND Ω_E

Now consider non-null optimal path $\Pi^*(C)$ which starts at point

$$\mathbf{x}^i = (x_0, x^i) \in \mathscr{X}^*, \qquad \mathscr{X}^* \triangleq X^* \times \{x_0\},$$

is generated by (p^*, e^*), and is given by

$$\mathbf{x}^*: k \to \mathbf{x}^k = \mathbf{x}^*(k), \qquad k \in \{i, i+1, \ldots, K\},$$

$\mathbf{x}^* = (\mathbf{x}_0^*, \mathbf{x}^*) = (\mathbf{x}_0^*, \mathbf{x}_1^*, \ldots, \mathbf{x}_n^*)$.

We shall prove first that

$$\mathbf{x}^*(j) \in \mathscr{X}^* \qquad \forall j \in \{i, i+1, \ldots, K\} \tag{4.16}$$

Indeed, by the definition of an optimal path, $\Pi^*(C)$ reaches Θ at stage K, and hence $\mathbf{x}^*(K) \in \Theta \subset \mathscr{X}^*$. If we suppose that $\mathbf{x}^*(j) \notin \mathscr{X}^*$ for $j < K$, then $p^*(\mathbf{x}^*(j))$, $e^*(\mathbf{x}^*(j))$ are not defined, and hence $\mathbf{x}^*(k)$ is not defined for $k > j$. Since $\mathbf{x}^*(k) \notin \Theta$ for $k \leqslant j$, we arrive at a contradiction, and so (4.16) is established.

It follows from (4.16) that

$$\Pi^*(C) \subset \mathscr{X}^* \tag{4.17}$$

Finally, it follows from (4.17) and from Corollary 2.1 that $\Pi^*(C)$ belongs to the game surface through $\mathbf{x}^i$. Of course, according to the definition of $\Pi^*(C)$ and of a game surface, the game surface through $\mathbf{x}^i$ is $\Sigma(C)$.

Now let us consider non-null paths $\Pi_E^{jl}(C')$ and $\Pi_P^{jm}(C'')$ which emanate from $\mathbf{x}^*(j)$, $j < K$, and which are generated by strategy pairs (p, e^*) and (p^*, e), respectively. Let us represent $\Pi_E^{jl}(C')$ and $\Pi_P^{jm}(C'')$ by

$$\mathbf{x}_E: k \to \mathbf{x}^k = \mathbf{x}_E(k), \qquad k \in \{j, j+1, \ldots, l\},$$

$$\text{and } \mathbf{x}_P: k \to \mathbf{x}^k = \mathbf{x}_P(k), \qquad k \in \{j, j+1, \ldots, m\},$$

respectively.

As a consequence of Lemmas 2.1 and 2.2, $\Pi_P^{jm}(C'')$ has no A-point and $\Pi_E^{jl}(C')$ has no B-point relative to $\Sigma(C)$.

Since $\mathbf{x}^*(j)$ belongs to $\mathscr{X}^*$ and $j < K$, points $\mathbf{x}_P(j+1)$ and $\mathbf{x}_E(j+1)$ of paths $\Pi_P^{jm}(C'')$ and $\Pi_E^{jl}(C')$, respectively, are defined. From the properties of these paths, that is from Lemmas 2.1 and 2.2, we deduce that

$$\mathbf{x}_P(j+1) \notin A/\Sigma_{j+1}(C) \tag{4.18}$$

$$\mathbf{x}_E(j+1) \notin B/\Sigma_{j+1}(C) \tag{4.19}$$

Next let us define sets $\Omega_P(j+1)$ and $\Omega_E(j+1)$, $j = i, i+1, \ldots, K-1$, by

Definition 4.4.

$$\Omega_P(j+1) \triangleq \{\mathbf{x}_P(j+1) \quad \text{for all strategy pairs } (p^*, e)\} \tag{4.20}$$

$$\Omega_E(j+1) \triangleq \{\mathbf{x}_E(j+1) \quad \text{for all strategy pairs } (p, e^*)\} \tag{4.21}$$

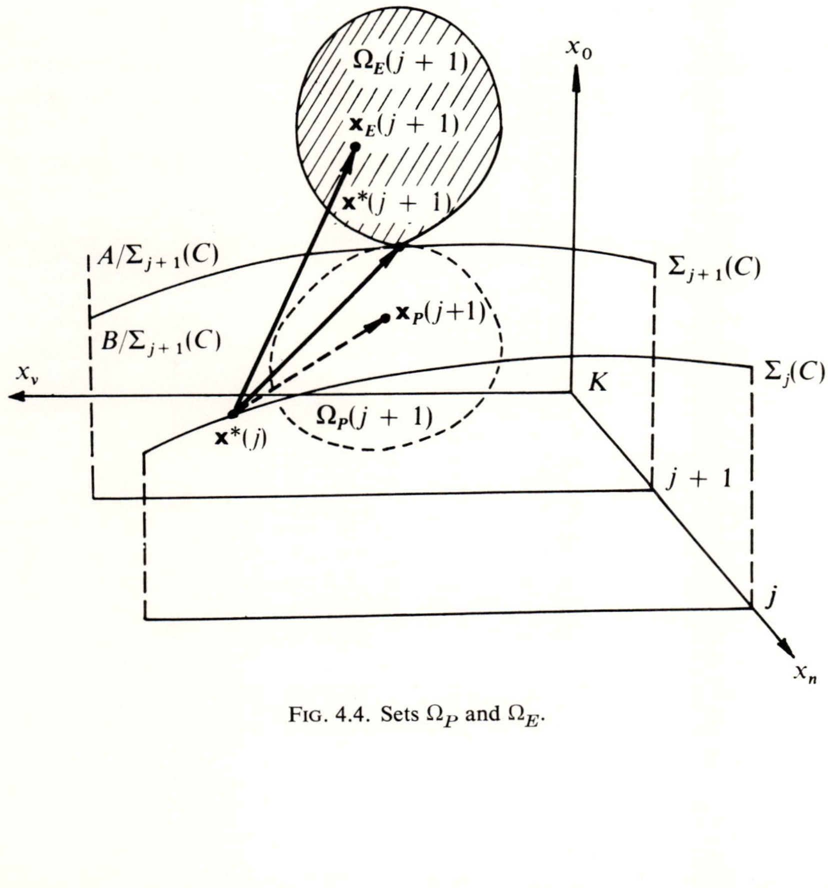

FIG. 4.4. Sets Ω_P and Ω_E.

From (4.8) and (4.9) we have

$$\Omega_P(j+1) = \{\mathbf{x}(j+1): \quad \mathbf{x}(j+1) = \mathbf{x}^*(j) + \mathbf{f}^j(\mathbf{x}^*(j), \mathbf{u}^*(j), \mathbf{v}(j)) \qquad \forall \mathbf{v}(j) \in K_v(\mathbf{x}^*(j))\} \tag{4.22}$$

$$\Omega_E(j+1) = \{\mathbf{x}(j+1): \quad \mathbf{x}(j+1) = \mathbf{x}^*(j) + \mathbf{f}^j(\mathbf{x}^*(j), \mathbf{u}(j), \mathbf{v}^*(j)) \qquad \forall \mathbf{u}(j) \in K_u(\mathbf{x}^*(j))\} \tag{4.23}$$

and from (4.18) and (4.19), since j is arbitrary, it follows that

$$\Omega_P(j+1) \cap (A/\Sigma_{j+1}(C)) = \varnothing \tag{4.24}$$

and

$$\Omega_E(j+1) \cap (B/\Sigma_{j+1}(C)) = \varnothing \tag{4.25}$$

for all $j \in \{i, i+1, \ldots, K-1\}$. Of course,

$$\mathbf{x}^*(j+1) \in \Omega_P(j+1) \cap \Omega_E(j+1)$$

4.8 x_0-DIRECTIONAL CONVEXITY

Definition 4.5. A set S in E^{n+1} is directionally convex with respect to vector w^0, or equivalently it is x_0^+*-directionally convex*, if for all $\mathbf{x}^1, \mathbf{x}^2 \in S$ and all $\nu \in [0, 1]$ there exists a κ, $-\infty < \kappa \leqslant 0$, such that

$$\mathbf{x}^1 + \nu(\mathbf{x}^2 - \mathbf{x}^1) + \kappa w^0 \in S$$

where $w^0 = (1, 0, \ldots, 0)$ is a unit vector in the x_0-direction.

S is x_0^-*-directionally convex*, if it is directionally convex with respect to $-w^0 = (-1, 0, \ldots, 0)$.

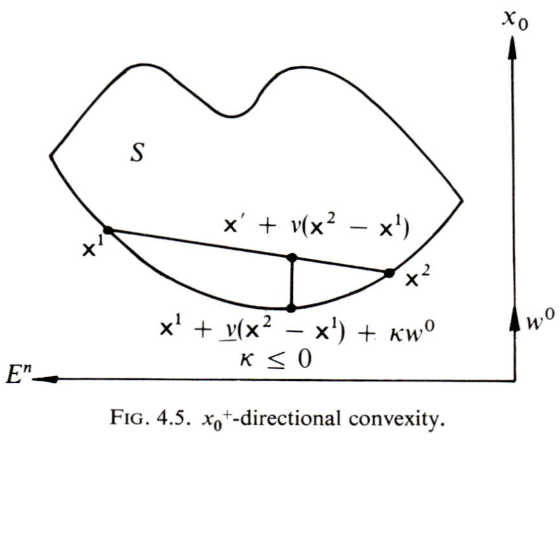

FIG. 4.5. x_0^+-directional convexity.

Clearly, directional convexity is weaker than convexity, since a convex set in E^{n+1} is directionally convex with respect to w^0 and with respect to any other unit vector in E^{n+1}, whereas the converse need not be true.

Next we introduce

Assumption 4.2. Set $\Omega_E(j+1)$ is x_0^+-directionally convex, and set $\Omega_P(j+1)$ is x_0^--directionally convex, for $j = i, i+1, \ldots, K-1$.

4.9 REGULAR POINTS OF A SURFACE $\Sigma_k(C)$

Definition 4.6. A point $\mathbf{x}^k$ of a surface $\Sigma_k(C)$ will be called a *regular point of* $\Sigma_k(C)$, if there exists an n-dimensional ball $\Delta(\mathbf{x}_k)$ in $\mathscr{X}_k$, with center $\mathbf{x}^k$, such that every point $\hat{\mathbf{x}}^k$ of $\Sigma_k(C)$ in $\Delta(\mathbf{x}_k)$ can be represented by

$$\hat{\mathbf{x}}^k = \mathbf{x}^k + \epsilon\eta^k + o(\epsilon)$$

where $\lim_{\epsilon\to 0} [\|o(\epsilon)\|/\epsilon] = 0$ and η^k belongs to an $(n-1)$-dimensional subspace of $E_k \times \{x_0\}$, the tangent plane of $\Sigma_k(C)$ at $\mathbf{x}^k$.

Now let us consider point $\mathbf{x}^*(j+1)$ of optimal path $\Pi^*(C)$ defined in Section 4.7. As shown in Section 4.7, path $\Pi^*(C)$ belongs to $\Sigma(C)$, and hence point $\mathbf{x}^*(j+1)$ belongs to $\Sigma_{j+1}(C)$.†

If grad $\Phi(\mathbf{x}^{j+1})$ is defined in a neighborhood of point $\mathbf{x}^*(j+1)$, where $\mathbf{x}^{j+1} \triangleq (x_0, x_1, \ldots, x_{n-1}, j+1)$ and

$$\text{grad}\,\Phi(\mathbf{x}^{j+1}) \triangleq \left(1, \frac{\partial V^*(x^{j+1})}{\partial x_1}, \ldots, \frac{\partial V^*(x^{j+1})}{\partial x_{n-1}}, 0\right)$$

then $\mathbf{x}^*(j+1)$ is a regular point of $\Sigma_{j+1}(C)$.

Let

$$\mathbf{x}_P(j+1) = \mathbf{x}^*(j+1) + \delta^P\mathbf{x}$$

$$\mathbf{x}_E(j+1) = \mathbf{x}^*(j+1) + \delta^E\mathbf{x}$$

where

$$\delta^P\mathbf{x} \triangleq \mathbf{f}^j(\mathbf{x}^*(j), \mathbf{u}^*(j), \mathbf{v}(j)) - \mathbf{f}^j(\mathbf{x}^*(j), \mathbf{u}^*(j), \mathbf{v}^*(j)) \tag{4.26}$$

$$\delta^E\mathbf{x} \triangleq \mathbf{f}^j(\mathbf{x}^*(j), \mathbf{u}(j), \mathbf{v}^*(j)) - \mathbf{f}^j(\mathbf{x}^*(j), \mathbf{u}^*(j), \mathbf{v}^*(j)) \tag{4.27}$$

Since $\mathbf{x}^*(j+1)$ and $\mathbf{x}_P(j+1)$ belong to $\Omega_P(j+1)$, and since $\Omega_P(j+1)$ is assumed to be x_0^--directionally convex, then, for all $\nu \in [0, 1]$,

† Since all points of $\Sigma_{j+1}(C)$ are interior points of $\Sigma_{j+1}(C)$, point $\mathbf{x}^*(j+1)$ is an interior point of $\Sigma_{j+1}(C)$.

here exists a κ, $0 \leqslant \kappa < +\infty$, such that

$$\mathbf{x}^*(j+1) + \nu\delta^P\mathbf{x} + \kappa w^0 \in \Omega_P(j+1)$$

Since $\Omega_P(j+1) \cap (A/\Sigma_{j+1}(C)) = \varnothing$, we have

$$\mathbf{x}^*(j+1) + \nu\,\delta^P\mathbf{x} + \kappa w^0 \notin A/\Sigma_{j+1}(C)$$

which, in turn, implies that

$$\mathbf{x}^*(j+1) + \nu\,\delta^P\mathbf{x} \notin A/\Sigma_{j+1}(C) \qquad \text{for all} \qquad \nu \in [0, 1] \tag{4.28}$$

Since $\mathbf{x}^*(j+1)$ is an interior point of $\Sigma_{j+1}(C)$, there exists $\sigma > 0$ such that, for all $\epsilon \in (0, \sigma)$,

$$\mathbf{x}^*(j+1) + \epsilon\,\delta^P\mathbf{x} \in \mathscr{X}_k$$

Then (4.28) implies that

$$\mathbf{x}^*(j+1) + \epsilon\,\delta^P\mathbf{x} \in \widetilde{B/\Sigma}_{j+1}(C) \tag{4.29}$$

where $\widetilde{B/\Sigma}_{j+1}(C) \triangleq (B/\Sigma_{j+1}(C)) \cup \Sigma_{j+1}(C)$

It follows from (4.29) with (4.13) and (4.15) that

$$\Phi(\mathbf{x}^*(j+1) + \epsilon\,\delta^P\mathbf{x}) \leqslant C \tag{4.30}$$

If grad $\Phi(\mathbf{x}^{j+1})$ is defined in a neighborhood point $\mathbf{x}^*(j+1)$, (4.30) can be rewritten as

$$\Phi(\mathbf{x}^*(j+1)) + \operatorname{grad}\Phi(\mathbf{x}^*(j+1)) \cdot \epsilon\,\delta^P\mathbf{x} + o(\epsilon) \leqslant C \tag{4.31}$$

where $\lim_{\epsilon\to 0}\,[o(\epsilon)/\epsilon] = 0$.

Since $\mathbf{x}^*(j+1)$ belongs to $\Sigma_{j+1}(C)$, and accordingly $\Phi(\mathbf{x}^*(j+1)) = C$, (4.31) becomes

$$\operatorname{grad}\Phi(\mathbf{x}^*(j+1)) \cdot \epsilon\,\delta^P\mathbf{x} + o(\epsilon) \leqslant 0 \tag{4.32}$$

Dividing by ϵ which is positive, we deduce from (4.32) that

$$\operatorname{grad}\Phi(\mathbf{x}^*(j+1)) \cdot \delta^P\mathbf{x} + o(\epsilon)/\epsilon \leqslant 0$$

Then, upon letting $\epsilon \to 0$, we obtain

$$\operatorname{grad}\Phi(\mathbf{x}^*(j+1)) \cdot \delta^P\mathbf{x} \leqslant 0$$

that is

$$\operatorname{grad}\Phi(\mathbf{x}^*(j+1)) \cdot [\mathbf{f}^j(\mathbf{x}^*(j), \mathbf{u}^*(j), \mathbf{v}(j)) - \mathbf{f}^j(\mathbf{x}^*(j), \mathbf{u}^*(j), \mathbf{v}^*(j))] \leqslant 0 \tag{4.33}$$

This relation holds for all $\mathbf{v}(j) \in K_v(\mathbf{x}^*(j))$.

Finally, (4.33) can be rewritten as

$$\text{grad}\,\Phi(\mathbf{x}^*(j+1))\cdot \mathbf{f}^j(\mathbf{x}^*(j),\mathbf{u}^*(j),\mathbf{v}(j)) \leqslant \text{grad}\,\Phi(\mathbf{x}^*(j+1))\cdot \mathbf{f}^j(\mathbf{x}^*(j),\mathbf{u}^*(j),\mathbf{v}^*(j)) \quad \forall \mathbf{v}(j)\in K_v(\mathbf{x}^*(j)) \tag{4.34}$$

By similar arguments based on the fact that $\Omega_E(j+1)$ is x_0^+-directionally convex and that $\Omega_E(j+1)\cap(B/\Sigma_{j+1}(C))=\varnothing$, one obtains

$$\text{grad}\,\Phi(\mathbf{x}^*(j+1))\cdot \mathbf{f}^j(\mathbf{x}^*(j),\mathbf{u}(j),\mathbf{v}^*(j)) \geqslant \text{grad}\,\Phi(\mathbf{x}^*(j+1))\cdot \mathbf{f}^j(\mathbf{x}^*(j),\mathbf{u}^*(j),\mathbf{v}^*(j)) \quad \forall \mathbf{u}(j)\in K_u(\mathbf{x}^*(j)) \tag{4.35}$$

4.10 REGULAR OPTIMAL PATHS

Definition 4.7. An optimal path $\Pi^*(C)$ given by $\mathbf{x}^*\colon\ k\to\mathbf{x}^k=\mathbf{x}^*(k)$, $k\in\{i,i+1,\ldots,K\}$, is called a regular optimal path, if $\mathbf{x}^*(k)$ is a regular point of $\Sigma_k(C)$ for all $k\in\{i,i+1,\ldots,K-1\}$.

Recall that all points of $\Sigma_k(C)$ are interior points of $\Sigma_k(C)$, and hence that $\mathbf{x}^*(k)$ is an interior point of $\Sigma_k(C)$, $k=i,i+1,\ldots,K$.

Let $\mathbf{x}^*\colon\ k\to x^k=\mathbf{x}^*(k)$, $k\in\{i,i+1,\ldots,K\}$, represent the projection π^* of $\Pi^*(C)$ on X^*. Since X_k is open in E_k, $\mathbf{x}^*(k)$ is an interior point of X_k, $k=i,i+1,\ldots,K$.

Next we introduce

Assumption 4.3. At every point $\mathbf{x}^*(k)$, $k=i,\ i+1,\ldots,K-1$, there exists a neighborhood $\Delta(\mathbf{x}^*(k))$ in X_k of $\mathbf{x}^*(k)$ on which p^* and e^* are of class C^1.

Since strategy pair (p^*,e^*) is optimal on X, it is optimal on $\Delta(\mathbf{x}^*(k))$, $k=i,i+1,\ldots,K-1$.

From (4.4) and Definition 2.3 we have

$$V^*(x^k)=\sum_{\nu=k}^{K-1} f_0^\nu(\mathbf{x}(\nu),p^*(\mathbf{x}(\nu)),e^*(\mathbf{x}(\nu))) \qquad \forall x^k\in\Delta(\mathbf{x}^*(k)) \tag{4.36}$$

where $\mathbf{x}\colon\ \nu\to x^\nu=\mathbf{x}(\nu)$ is the solution of difference equation

$$\mathbf{x}(\nu+1)-\mathbf{x}(\nu)=f^\nu(\mathbf{x}(\nu),p^*(\mathbf{x}(\nu)),e^*(\mathbf{x}(\nu)))$$

with initial condition $\mathbf{x}(k)=x^k$, defined on $\{k,k+1,\ldots,K\}$.

Then we have

Lemma 4.1. *At points* $\mathbf{x}^*(k)$, $k = i, i+1, \ldots, K$, *of optimal path* $\Pi^*(C)$ *on game surface* $\Sigma(C)$ *there exists a neighborhood in* E_k *on which* grad $\Phi(\mathbf{x}^k)$ *is defined,*† *where*

$$\operatorname{grad} \Phi(\mathbf{x}^k) \triangleq \left(1, \frac{\partial V^*(x^k)}{\partial x_1}, \ldots, \frac{\partial V^*(x^k)}{\partial x_{n-1}}, 0\right)$$

For $k = i, i+1, \ldots, K-1$, Lemma 4.1 is a direct consequence of Assumption 4.3, together with (4.36) and with the fact that the functions f_0^ν, $\nu = k, k+1, \ldots, K-1$, are of class C^1 on $\mathscr{D} \times U \times V$.

For $k = K$, it follows from (4.5) that $V^*(x^K) = 0$ for all $x^K \in \theta$, and hence

$$\operatorname{grad} \Phi(\mathbf{x}^K) = (1, 0, \ldots, 0)$$

4.11 VARIATIONAL DIFFERENCE EQUATION

Again let us consider points $\mathbf{x}^*(k)$, $k = i, i+1, \ldots, K$, of optimal path $\Pi^*(C)$, and let us introduce

Assumption 4.4. Matrix M^k given by

$$M^k \triangleq \left[I + \frac{d\mathsf{f}^k(\mathsf{x}^k, p^*(x^k), e^*(x^k))}{d\mathsf{x}}\bigg|_{\mathsf{x}^k = \mathbf{x}^*(k)}\right]$$

where

$$\frac{d\mathsf{f}^k(\mathsf{x}^k, p^*(x^k), e^*(x^k))}{d\mathsf{x}} \triangleq \frac{d\mathsf{f}^k}{d\mathsf{x}} \triangleq \frac{\partial \mathsf{f}^k}{\partial \mathsf{x}} + \frac{\partial \mathsf{f}^k}{\partial u}\frac{\partial p^*}{\partial \mathsf{x}} + \frac{\partial \mathsf{f}^k}{\partial v}\frac{\partial e^*}{\partial \mathsf{x}}$$

and I is the $(n+1) \times (n \times 1)$ unit matrix, is defined and nonsingular for $k = i, i+1, \ldots, K-1$.

$$\frac{\partial \mathsf{f}^k}{\partial \mathsf{x}} \triangleq \left[\frac{\partial f_\nu^k(\mathsf{x}^k, u, v)}{\partial x_\alpha}\right] \qquad \nu, \alpha = 0, 1, \ldots, n$$

$$\frac{\partial p^*}{\partial \mathsf{x}} \triangleq \left[\frac{\partial p_\nu^*(x^k)}{\partial x_\alpha}\right] \qquad \begin{aligned} \nu &= 1, 2, \ldots, r \\ \alpha &= 0, 1, \ldots, n \end{aligned}$$

$$\frac{\partial e^*}{\partial \mathsf{x}} \triangleq \left[\frac{\partial e_\nu^*(x^k)}{\partial x_\alpha}\right] \qquad \begin{aligned} \nu &= 1, 2, \ldots, s \\ \alpha &= 0, 1, \ldots, n \end{aligned}$$

† Recall that $\mathbf{x}^k \triangleq (x_0, x_1, \ldots, x_{n-1}, k)$.

evaluated at $\mathrm{x}^k = \mathbf{x}^*(k), u = p^*(\mathbf{x}^*(k)), v = e^*(\mathbf{x}^*(k)), k = i, i+1, \ldots, K-1$.

Now let us consider the bound vector

$$\eta^k \triangleq (\eta_0{}^k, \eta_1{}^k, \ldots, \eta_n{}^k) \in E^{n+1}$$

at point x^k.

Let $\boldsymbol{\eta}$: $k \rightarrow \eta^k = \boldsymbol{\eta}(k)$, $k \in \{i, i+1, \ldots, K\}$, be defined in the following manner:

1. Let $\eta^i = \boldsymbol{\eta}(i)$ be given at the initial point $\mathbf{x}^*(i)$ of $\Pi^*(C)$. Then

$$\hat{\mathrm{x}}^i \triangleq \mathbf{x}(i) = \mathbf{x}^*(i) + \epsilon\boldsymbol{\eta}(i) + o(\epsilon)$$

is a point in a neighborhood in $\mathscr{X}_i$ of $\mathbf{x}^*(i)$, where $\lim_{\epsilon\to 0} [|o(\epsilon)|/\epsilon] = 0$.

2. Let $\hat{\mathrm{x}}^i$ be transformed by the state equation (4.7) with the same strategy pair (p^*, e^*) that generates $\Pi^*(C)$.

The state equation can be written

$$\mathbf{x}(k+1) - \mathbf{x}(k) = \mathrm{f}^k(\mathbf{x}(k), p^*(\mathbf{x}(k)), e^*(\mathbf{x}(k))) \tag{4.37}$$

for $k = i, i+1, \ldots, K-1$

One readily deduces from the state equation that

$$\mathbf{x}(i+1) = \mathbf{x}^*(i+1) + \epsilon\left[I + \left.\frac{d\mathrm{f}^i(\mathrm{x}^i, p^*(x^i), e^*(x^i))}{d\mathrm{x}}\right|_{\mathrm{x}^i=\mathbf{x}^*(i)}\right]\boldsymbol{\eta}(i) + o(\epsilon) \tag{4.38}$$

that is

$$\mathbf{x}(i+1) = \mathbf{x}^*(i+1) + \epsilon M^i\boldsymbol{\eta}(i) + o(\epsilon) \tag{4.39}$$

Letting

$$\boldsymbol{\eta}(i+1) = \eta^{i+1} \triangleq M^i\boldsymbol{\eta}(i)$$

we have

$$\mathbf{x}(i+1) = \mathbf{x}^*(i+1) + \epsilon\boldsymbol{\eta}(i+1) + o(\epsilon) \tag{4.40}$$

Likewise, if

$$\mathbf{x}(k) = \mathbf{x}^*(k) + \epsilon\boldsymbol{\eta}(k) + o(\epsilon) \tag{4.41}$$

then

$$\begin{aligned}\mathbf{x}(k+1) &= \mathbf{x}^*(k+1) \\ &\quad + \epsilon\left[I + \frac{d\mathbf{f}^k(\mathbf{x}^k, p^*(x^k), e^*(x^k))}{d\mathbf{x}}\bigg|_{\mathbf{x}^k=\mathbf{x}^*(k)}\right]\boldsymbol{\eta}(k) + o(\epsilon) \\ &= \mathbf{x}^*(k+1) + \epsilon M^k\boldsymbol{\eta}(k) + o(\epsilon)\end{aligned}$$

Letting

$$\boldsymbol{\eta}(k+1) = \eta^{k+1} \triangleq M^k\boldsymbol{\eta}(k)$$

we have

$$\mathbf{x}(k+1) = \mathbf{x}^*(k+1) + \epsilon\boldsymbol{\eta}(k+1) + o(\epsilon) \tag{4.42}$$

and so

$$\boldsymbol{\eta}(k+1) - \boldsymbol{\eta}(k) = \left[\frac{d\mathbf{f}^k(\mathbf{x}^k, p^*(x^k), e^*(x^k)}{d\mathbf{x}}\bigg|_{\mathbf{x}^k=\mathbf{x}^*(k)}\right]\boldsymbol{\eta}(k) \tag{4.43}$$

for $k = i, i+1, \ldots, K-1$.

Equation (4.43) is the *variational difference equation.*

Note that $f_n^k(x, u, v) = 1$, $k = i, i+1, \ldots, K-1$, and hence

$$\boldsymbol{\eta}_n(k+1) - \boldsymbol{\eta}_n(k) = 0 \tag{4.44}$$

It follows that

$$\boldsymbol{\eta}_n(i) = \boldsymbol{\eta}_n(i+1) = \cdots = \boldsymbol{\eta}_n(K) \tag{4.45}$$

4.12 LINEAR TRANSFORMATION AND INVERSE TRANSFORMATION

For given $\boldsymbol{\eta}(i) = \eta^i$ we have

$$\boldsymbol{\eta}(k+1) = M^k\boldsymbol{\eta}(k) = M^k(M^{k-1}\boldsymbol{\eta}(k-1)) = M^kM^{k-1}\cdots M^i\boldsymbol{\eta}(i)$$

Let

$$\mathscr{A}^k \triangleq M^kM^{k-1}\cdots M^i$$

for $k = i, i+1, \ldots, K-1$.

Since matrices M^k, $k = i, i+1, \ldots, K-1$, are assumed nonsingular, matrices $\mathscr{A}^k$, $k = i, i+1, \ldots, K-1$, are nonsingular. Hence the linear transformation

$$\boldsymbol{\eta}(k+1) = \mathscr{A}^k\boldsymbol{\eta}(i), \qquad k = i, i+1, \ldots, K-1 \tag{4.46}$$

is nonsingular; that is, the inverse of $\mathscr{A}^k$ exists, and we have

$$\boldsymbol{\eta}(i) = (\mathscr{A}^k)^{-1}\boldsymbol{\eta}(k+1)$$

If $\boldsymbol{\eta}(K) = \eta^K$ is given, then

$$\boldsymbol{\eta}(i) = (\mathscr{A}^{K-1})^{-1}\boldsymbol{\eta}(K)$$

and

$$\boldsymbol{\eta}(k) = \mathscr{A}^{k-1}\boldsymbol{\eta}(i) = \mathscr{A}^{k-1}(\mathscr{A}^{K-1})^{-1}\boldsymbol{\eta}(K)$$

where $\mathscr{A}^{i-1} \triangleq I$.

Let

$$\mathscr{B}^k \triangleq \mathscr{A}^{k-1}(\mathscr{A}^{K-1})^{-1}$$

where $\mathscr{B}^i \triangleq (\mathscr{A}^{K-1})^{-1}$.

Then we have the inverse linear transformation

$$\boldsymbol{\eta}(k) = \mathscr{B}^k\boldsymbol{\eta}(K) \tag{4.47}$$

for $k = i, i+1, \ldots, K-1$.

4.13 SOME PROPERTIES OF LINEAR TRANSFORMATIONS $\mathscr{A}^k$ AND $\mathscr{B}^k$

Now let $P(\mathbf{x}^*(i))$ be a given γ-dimensional hyperplane, $1 \leqslant \gamma \leqslant n$, containing point $\mathbf{x}^*(i)$ of $\Pi^*(C)$. We shall be interested in the transform of plane $P(\mathbf{x}^*(i))$ due to linear transformation $\mathscr{A}^{k-1}$, namely,

$$\begin{aligned} P(\mathbf{x}^*(k)) &= \mathscr{A}^{k-1}P(\mathbf{x}^*(i)) \\ &\triangleq \{\mathbf{x}^*(k) + \mathscr{A}^{k-1}\eta^i : \quad \mathbf{x}^*(i) + \eta^i \in P(\mathbf{x}^*(i))\} \end{aligned}$$

Lemma 4.2. *The transform $P(\mathbf{x}^*(k))$, due to linear transformation $\mathscr{A}^{k-1}$, of plane $P(\mathbf{x}^*(i))$ containing point $\mathbf{x}^*(i)$ of optimal path $\Pi^*(C)$ has these properties:*

(a) *$P(\mathbf{x}^*(k))$ is defined for $k = i, i+1, \ldots, K$;*

(b) *$P(\mathbf{x}^*(k))$ is a plane of the same dimension as $P(\mathbf{x}^*(i))$.*

Since $P(\mathbf{x}^*(i))$ is a γ-dimensional plane, there exists a basis in $P(\mathbf{x}^*(i))$, that is γ linearly independent vectors $e_\nu(i)$, $\nu = 1, 2, \ldots, \gamma$, such that any nonzero vector η^i in $P(\mathbf{x}^*(i))$ is of the form

$$\eta^i = \boldsymbol{\eta}(i) = \sum_{\nu=1}^{\gamma} c_\nu e_\nu(i)$$

where the c_ν are constants, not all of which are zero.

Let $e_\nu(i)$, $\nu = 1, 2, \ldots, \gamma$, be transformed along $\Pi^*(C)$ by equation (4.46); then

$$e_\nu(k) = \mathscr{A}^{k-1} e_\nu(i) \qquad \nu = 1, 2, \ldots, \gamma$$

Since the vectors $e_\nu(i)$, $\nu = 1, 2, \ldots, \gamma$, are linearly independent and $\mathscr{A}^{k-1}$ is nonsingular, the vectors $e_\nu(k)$, $\nu = 1, 2, \ldots, \gamma$, are also linearly independent.

Any vector η^k of $P(\mathbf{x}^*(k))$ is of the form

$$\eta^k = \boldsymbol{\eta}(k) = \mathscr{A}^{k-1}\boldsymbol{\eta}(i) = \mathscr{A}^{k-1}\left(\sum_{\nu=1}^{\gamma} c_\nu e_\nu(i)\right)$$

and since $\mathscr{A}^{k-1}$ is linear,

$$\eta^k = \sum_{\nu=1}^{\gamma} c_\nu \mathscr{A}^{k-1} e_\nu(i) = \sum_{\nu=1}^{\gamma} c_\nu e_\nu(k)$$

It follows that $P(\mathbf{x}^*(k))$, $k = i, i + 1, \ldots, K$, is defined and its dimension is γ, and so Lemma 4.2 is established.

By similar arguments, using the properties of inverse transformation $\mathscr{B}^k$, one can prove

Lemma 4.3. *The transform $P(\mathbf{x}^*(k))$, due to linear transformation $\mathscr{B}^k$, of a plane $P(\mathbf{x}^*(K))$ containing point $\mathbf{x}^*(K)$ of optimal path $\Pi^*(C)$ has these properties:*

(a) *$P(\mathbf{x}^*(k))$ is defined for $k = i, i + 1, \ldots, K$;*

(b) *$P(\mathbf{x}^*(k))$ is a plane of the same dimension as $P(\mathbf{x}^*(K))$.*

4.14 TRANSFORMATION OF A TANGENT PLANE

Since, according to Lemma 4.1, grad $\Phi(\mathrm{x}^k)$ is defined at all points $\mathbf{x}^*(k)$, $k = i, i + 1, \ldots, K - 1$, $\Pi^*(C)$ is a regular optimal path on $\Sigma(C)$. Consider a neighborhood in $\Sigma_i(C)$ of $\mathbf{x}^*(i)$,† namely,

$$\Delta(\mathbf{x}^*(i)) \triangleq \{\mathrm{x}^i\colon \quad \mathrm{x}^i = \mathbf{x}(i) \in \Sigma_i(C)\}$$

where

$$\mathbf{x}(i) = \mathbf{x}^*(i) + \epsilon\eta^i + o(\epsilon)$$

and

1. $\lim_{\epsilon\to 0} [\|o(\epsilon)\|/\epsilon] = 0$.

2. $\eta^i \in P_{\Sigma_i}(\mathbf{x}^*(i))$, the tangent plane of $\Sigma_i(C)$ at $\mathbf{x}^*(i)$, which is an $(n - 1)$-dimensional plane in $E_i \times \{x_0\}$.

† Recall that $\mathbf{x}^*(i)$ is an interior point of $\Sigma_i(C)$.

We shall be interested in a transformation of $\Delta(\mathbf{x}^*(i))$ from stage i to stage k by means of state equation (4.37); that is, with a point $\mathbf{x}^i = (x_0, x^i)$ of $\Delta(\mathbf{x}^*(i))$, we shall associate optimal path $\hat{\Pi}^*(C)$ which emanates from $\mathbf{x}^i$ and which is generated by the same strategy pair (p^*, e^*) that generates $\Pi^*(C)$ emanating from $\mathbf{x}^*(i)$. Then the transform of $\Delta(\mathbf{x}^*(i))$ is

$$\Delta(\mathbf{x}^*(k)) \triangleq \{\mathbf{x}^k = \mathbf{x}(k): \quad \mathbf{x}(k) \in \hat{\Pi}^*(C) \qquad \forall \mathbf{x}^i \in \Delta(\mathbf{x}^*(i))\}$$

From (4.42) we have

$$\Delta(\mathbf{x}^*(k)) = \{\mathbf{x}^k = \mathbf{x}(k): \quad \mathbf{x}(k) = \mathbf{x}^*(k) + \epsilon\boldsymbol{\eta}(k) + o(\epsilon)\}$$

where $\boldsymbol{\eta}$ is the solution of variational equation (4.43) with initial condition $\boldsymbol{\eta}(i) = \eta^i$; that is,

$$\boldsymbol{\eta}(k) = \mathscr{A}^{k-1}\eta^i$$

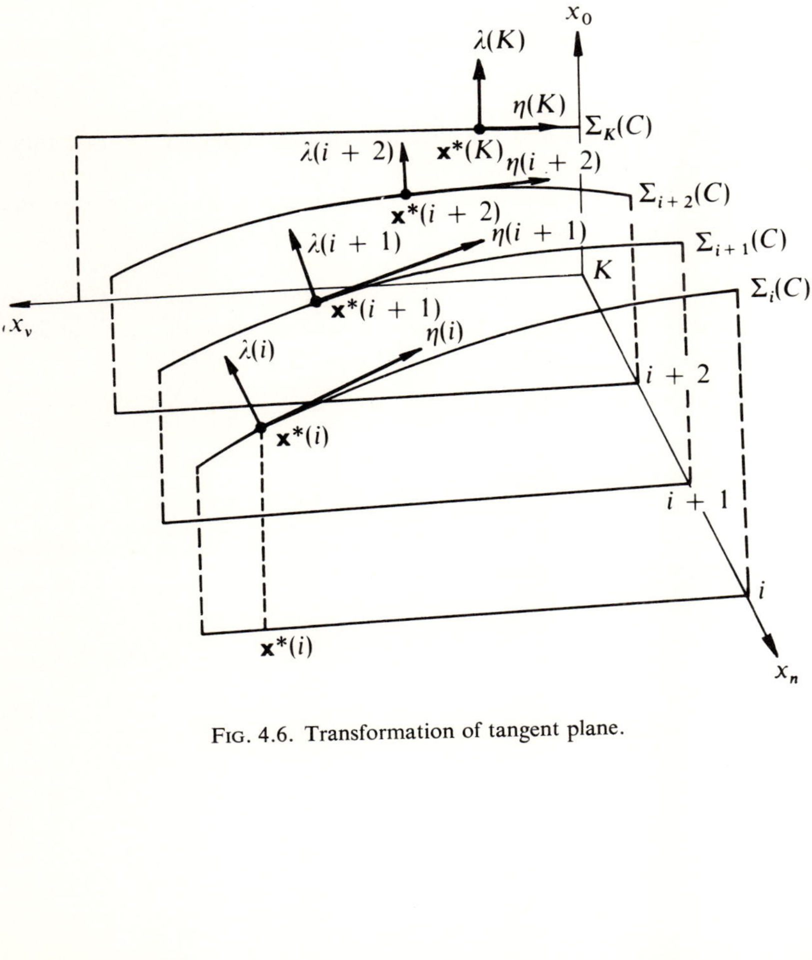

FIG. 4.6. Transformation of tangent plane.

Since condition (4.17) applies to all optimal paths, $\hat{\Pi}^*(C) \subset \mathscr{X}^*$ for all $\mathbf{x}^i \in \Delta(\mathbf{x}^*(i))$. It follows from Corollary 2.1 that

$$\mathbf{x}(k) = \mathbf{x}^*(k) + \epsilon\boldsymbol{\eta}(k) + o(\epsilon) \in \Sigma_k(C)$$

where $\lim_{\epsilon\to 0} [\|o(\epsilon)\|/\epsilon] = 0$

Hence,

$$\Phi(\mathbf{x}^*(k) + \epsilon\boldsymbol{\eta}(k) + o(\epsilon)) = C \tag{4.48}$$

Since $\mathbf{x}^*(k)$ belongs to $\Sigma_k(C)$, and accordingly $\Phi(\mathbf{x}^*(k)) = C$, (4.48) can be rewritten as

$$\operatorname{grad} \Phi(\mathbf{x}^*(k)) \cdot (\epsilon\boldsymbol{\eta}(k) + o(\epsilon)) + o(\|\epsilon\boldsymbol{\eta}(k) + o(\epsilon)\|) = 0 \tag{4.49}$$

Equation (4.49) can be written more simply as

$$\epsilon \operatorname{grad} \Phi(\mathbf{x}^*(k)) \cdot \boldsymbol{\eta}(k) + o(\epsilon) = 0$$

where $\lim_{\epsilon\to 0} [o(\epsilon)/\epsilon] = 0$

Upon dividing by ϵ and allowing $\epsilon \to 0$ we obtain

$$\operatorname{grad} \Phi(\mathbf{x}^*(k)) \cdot \boldsymbol{\eta}(k) = 0 \tag{4.50}$$

Furthermore, (4.45) and the fact that η^i belongs to $E_i \times \{x_0\}$ imply that $\boldsymbol{\eta}(k)$ belongs to $E_k \times \{x_0\}$.

Since $\eta^i \in P_{\Sigma_i}(\mathbf{x}^*(i))$ and since $\mathscr{A}^{k-1}$ is a linear nonsingular transformation, condition (4.50) leads to

Lemma 4.4. *If Assumptions* 4.3 *and* 4.4 *are satisfied, then the transform, due to linear transformation* $\mathscr{A}^{k-1}$, *of the tangent plane of* $\Sigma_i(C)$ *at* $\mathbf{x}^*(i)$ *is the tangent plane* $P_{\Sigma_k}(\mathbf{x}^*(k))$ *of* $\Sigma_k(C)$ *at* $\mathbf{x}^*(k)$; *that is,*

$$P_{\Sigma_k}(\mathbf{x}^*(k)) = \mathscr{A}^{k-1} P_{\Sigma_i}(\mathbf{x}^*(i)) \qquad \textit{for} \qquad k = i, i+1, \ldots, K$$

4.15 ADJOINT EQUATIONS

Next let us introduce $\boldsymbol{\lambda}(k) \in E^{n+1}$, $k = i, i+1, \ldots, K$, satisfying the equation adjoint to variational equation (4.43),

$$\boldsymbol{\lambda}(k+1) - \boldsymbol{\lambda}(k) = -\left[\frac{d\mathbf{f}^k(\mathbf{x}^k, p^*(x^k), e^*(x^k))}{d\mathbf{x}}\bigg|_{\mathbf{x}^k = \mathbf{x}^*(k)}\right]^{\mathrm{T}} \boldsymbol{\lambda}(k+1) \tag{4.51}$$

for $k = i, i+1, \ldots, K-1$.

This equation can be rewritten as

$$\boldsymbol{\lambda}(k) = \left[I + \frac{d\mathbf{f}^k(\mathbf{x}^k, p^*(x^k), e^*(x^k))}{d\mathbf{x}}\bigg|_{\mathbf{x}^k=\mathbf{x}^*(k)} \right]^{\mathrm{T}} \boldsymbol{\lambda}(k+1)$$

for $k = i, i+1, \ldots, K-1$.

Since matrices

$$\left[I + \frac{d\mathbf{f}^k(\mathbf{x}^k, p^*(x^k), e^*(x^k))}{d\mathbf{x}}\bigg|_{\mathbf{x}^k=\mathbf{x}^*(k)} \right], \qquad k = i, i+1, \ldots, K-1$$

are assumed to be nonsingular, so are matrices

$$\left[I + \frac{d\mathbf{f}^k(\mathbf{x}^k, p^*(x^k), e^*(x^k))}{d\mathbf{x}}\bigg|_{\mathbf{x}^k=\mathbf{x}^*(k)} \right]^{\mathrm{T}}.$$

Thus their inverses exist; that is

$$\boldsymbol{\lambda}(k+1) = \left\{ \left[I + \frac{d\mathbf{f}^k(\mathbf{x}^k, p^*(x^k), e^*(x^k))}{d\mathbf{x}}\bigg|_{\mathbf{x}^k=\mathbf{x}^*(k)} \right]^{\mathrm{T}} \right\}^{-1} \boldsymbol{\lambda}(k) \tag{4.52}$$

for $k = i, i+1, \ldots, K-1$. Hence, vectors $\boldsymbol{\lambda}(k)$, $k = i, i+1, \ldots, K$, are defined for any given $\boldsymbol{\lambda}(i)$ or $\boldsymbol{\lambda}(K)$.

It follows from Eq. (4.52) and variational equation (4.43) that

$$\sum_{\nu=0}^{n} \boldsymbol{\lambda}_\nu(i)\boldsymbol{\eta}_\nu(i) = \sum_{\nu=0}^{n} \boldsymbol{\lambda}_\nu(i+1)\boldsymbol{\eta}_\nu(i+1) = \cdots = \sum_{\nu=0}^{n} \boldsymbol{\lambda}_\nu(K)\boldsymbol{\eta}_\nu(K) \tag{4.53}$$

or equivalently

$$\boldsymbol{\lambda}(i) \cdot \boldsymbol{\eta}(i) = \boldsymbol{\lambda}(i+1) \cdot \boldsymbol{\eta}(i+1) = \cdots = \boldsymbol{\lambda}(K) \cdot \boldsymbol{\eta}(K) \tag{4.54}$$

Since x_0 and x_n do not occur in the argument of $\mathbf{f}^k(\mathbf{x}^k, p^*(x^k), e^*(x^k))$, $k = 0, 1, \ldots, K-1$, Eq. (4.51) implies that

$$\boldsymbol{\lambda}_0(k+1) - \boldsymbol{\lambda}_0(k) = 0, \qquad \boldsymbol{\lambda}_n(k+1) - \boldsymbol{\lambda}_n(k) = 0$$

for $k = i, i+1, \ldots, K-1$. It follows that

$$\boldsymbol{\lambda}_0(i) = \boldsymbol{\lambda}_0(i+1) = \cdots = \boldsymbol{\lambda}_0(K) \tag{4.55}$$

$$\boldsymbol{\lambda}_n(i) = \boldsymbol{\lambda}_n(i+1) = \cdots = \boldsymbol{\lambda}_n(K) \tag{4.56}$$

From (4.54) we can draw the conclusion that, if initial vectors $\lambda^i = \boldsymbol{\lambda}(i)$ and $\eta^i = \boldsymbol{\eta}(i)$ are chosen such that

$$\lambda^i \cdot \eta^i = 0 \qquad \text{with} \qquad \|\lambda^i\| \, \|\eta^i\| \neq 0$$

That is, if λ^i is perpendicular to η^i, then

$$\boldsymbol{\lambda}(k) \cdot \boldsymbol{\eta}(k) = 0 \qquad k = i, i+1, \ldots, K \tag{4.57}$$

Consequently $\boldsymbol{\lambda}(k)$ is perpendicular to $\boldsymbol{\eta}(k)$ at all points of $\Pi^*(C)$.†

Also, if $P(\mathbf{x}^*(i))$ is an $(n-1)$-dimensional plane containing point $\mathbf{x}^*(i)$ of $\Pi^*(C)$, and if $\lambda^i = \boldsymbol{\lambda}(i)$ is perpendicular to that plane, then $\boldsymbol{\lambda}(k)$ is perpendicular to the transform

$$P(\mathbf{x}^*(k)) = \mathscr{A}^{k-1}P(\mathbf{x}^*(i)) \tag{4.58}$$

of $P(\mathbf{x}^*(i))$ due to linear transformation $\mathscr{A}^{k-1}$, $k = i, i+1, \ldots, K$.

Finally, if $\lambda^i = \boldsymbol{\lambda}(i)$ is chosen perpendicular to the tangent plane $P_{\Sigma_i}(\mathbf{x}^*(i))$ of $\Sigma_i(C)$ at $\mathbf{x}^*(i)$, then it follows from (4.57) and Lemma 4.4 that $\boldsymbol{\lambda}(k)$ is perpendicular to the tangent plane $P_{\Sigma_k}(\mathbf{x}^*(k))$ of $\Sigma_k(C)$ at $\mathbf{x}^*(k)$, $k = i, i+1, \ldots, K$.

4.16 GRADIENT AND ADJOINT VECTOR

The x_0-component and the x_n-component of $\boldsymbol{\lambda}(k)$ are constants. Thus, if we choose $\boldsymbol{\lambda}_0(i) = 1$ and $\boldsymbol{\lambda}_n(i) = 0$, we have $\boldsymbol{\lambda}_0(k) = 1$ and $\boldsymbol{\lambda}_n(k) = 0$ for $k = i, i+1, \ldots, K$. Furthermore, if we choose $\boldsymbol{\lambda}(i)$ perpendicular to $P_{\Sigma_i}(\mathbf{x}^*(i))$, then, as pointed out above, $\boldsymbol{\lambda}(k)$ is perpendicular to $P_{\Sigma_k}(\mathbf{x}^*(k))$ for $k = i, i+1, \ldots, K$. Clearly, this choice of $\boldsymbol{\lambda}(i)$ implies that $\boldsymbol{\lambda}(i) = \operatorname{grad} \Phi(\mathrm{x}^i)$ at point $\mathrm{x}^i = \mathbf{x}^*(i)$.

Since both grad $\Phi(\mathrm{x}^k)$ at point $\mathrm{x}^k = \mathbf{x}^*(k)$ and $\boldsymbol{\lambda}(k)$

(i) belong to $E_k \times \{x_0\}$

(ii) have zeroth component equal to 1,

(iii) are perpendicular to $P_{\Sigma_k}(\mathbf{x}^*(k))$

for $k = i, i+1, \ldots, K$, it follows that

$$\boldsymbol{\lambda}(k) = \operatorname{grad} \Phi(\mathrm{x}^k) \qquad \text{at point} \qquad \mathrm{x}^k = \mathbf{x}^*(k) \tag{4.59}$$

for $k = i, i+1, \ldots, K$. And so we deduce from (4.34) that

$$\boldsymbol{\lambda}(k+1) \cdot \mathrm{f}^k(\mathbf{x}^*(k), \mathbf{u}^*(k), \mathbf{v}(k)) \leqslant \boldsymbol{\lambda}(k+1) \cdot \mathrm{f}^k(\mathbf{x}^*(k), \mathbf{u}^*(k), \mathbf{v}^*(k)) \quad \forall \mathbf{v}(k) \in K_v(\mathbf{x}^*(k)) \tag{4.60}$$

for $k = i, i+1, \ldots, K-1$, and from (4.35) that

$$\boldsymbol{\lambda}(k+1) \cdot \mathrm{f}^k(\mathbf{x}^*(k), \mathbf{u}(k), \mathbf{v}^*(k)) \geqslant \boldsymbol{\lambda}(k+1) \cdot \mathrm{f}^k(\mathbf{x}^*(k), \mathbf{u}^*(k), \mathbf{v}^*(k)) \quad \forall \mathbf{u}(k) \in K_u(\mathbf{x}^*(k)) \tag{4.61}$$

† Of course, $\|\boldsymbol{\lambda}(k)\| \, \|\boldsymbol{\eta}(k)\| \neq 0$.

for $k = i, i+1, \ldots, K-1$, where

$$\mathbf{x}^*(k) \triangleq (\mathbf{x}_0{}^*(k), \mathbf{x}^*(k)) \in \Pi^*(C)$$

and

$$\mathbf{u}^*(k) = p^*(\mathbf{x}^*(k)), \qquad \mathbf{v}^*(k) = e^*(\mathbf{x}^*(k))$$

4.17 TRANSVERSALITY CONDITION

At the terminal point $\mathbf{x}^*(K)$ of optimal path $\Pi^*(C)$

$$\boldsymbol{\lambda}(K) \cdot \boldsymbol{\eta}(K) = 0 \qquad \forall \boldsymbol{\eta}(K) \in P_{\Sigma_K}(\mathbf{x}^*(K))$$

The intersection of game surface $\Sigma(C)$ with Θ† is

$$\Theta \cap \Sigma(C) = \{\mathbf{x}\colon \ \Phi(\mathbf{x}) = C, \ x \in \theta\}$$

Because of (4.5), the condition $x \in \theta$ implies that

$$\Phi(\mathbf{x}) \triangleq x_0 + V^*(x) = x_0$$

Hence,

$$\Theta \cap \Sigma(C) = \{\mathbf{x}\colon \ x_0 = C, \ x \in \theta\}$$

Consequently, $\Theta \cap \Sigma(C)$ may be deduced from θ by translation parallel to the x_0-axis; and since $\theta \triangleq X_K \triangleq R \cap E_K$, $P_{\Sigma_K}(\mathbf{x}^*(K))$ is an $(n-1)$-dimensional plane perpendicular to the x_0-axis.

It follows that

$$\boldsymbol{\lambda}_1(K) = \cdots = \boldsymbol{\lambda}_{n-1}(K) = \boldsymbol{\lambda}_n(K) = 0 \tag{4.62}$$

at the terminal point of $\Pi^*(C)$. This is the transversality condition.

4.18 A MIN–MAX PRINCIPLE

Upon introducing the Hamiltonian

$$\mathscr{H}^{k+1}(\boldsymbol{\lambda}(k+1), \mathbf{x}(k), \mathbf{u}(k), \mathbf{v}(k)) \triangleq \boldsymbol{\lambda}(k+1) \cdot \mathbf{f}^k(\mathbf{x}(k), \mathbf{u}(k), \mathbf{v}(k))$$

for $k = i, i+1, \ldots, K-1$, the results of the preceding sections can be embodied in

Theorem 4.1.‡ *Let $\Pi^*(C)$ be an optimal path generated by strategies p^*, e^* and represented by $\mathbf{x}^*\colon \ k \to \mathbf{x}^k = \mathbf{x}^*(k)$, and let π^* be its projection*

† $\Theta \triangleq \theta \times \{x_0\}$.

‡ For necessary and sufficient conditions, see Reference 33.

on X^* *represented by* $\mathbf{x}^*\colon\ k \to x^k = \mathbf{x}^*(k),\ k \in \{i, i+1, \ldots, K\}$. *If Assumptions* 4.1–4.4 *are satisfied, then there exist nonzero vectors* $\boldsymbol{\lambda}(k)$, $k = i, i+1, \ldots, K$ *which satisfy adjoint equation* (4.51), *such that*

$$\text{(a)} \quad \min_{u \in K_u(\mathbf{x}^*(k))} \mathscr{H}^{k+1}(\boldsymbol{\lambda}(k+1), \mathbf{x}^*(k), u, e^*(\mathbf{x}^*(k)))$$

$$= \max_{v \in K_v(\mathbf{x}^*(k))} \mathscr{H}^{k+1}(\boldsymbol{\lambda}(k+1), \mathbf{x}^*(k), p^*(\mathbf{x}^*(k)), v)$$

$$= \mathscr{H}^{k+1}(\boldsymbol{\lambda}(k+1), \mathbf{x}^*(k), p^*(\mathbf{x}^*(k)), e^*(\mathbf{x}^*(k)))$$

for $k = i, i+1, \ldots, K-1$

$$\text{(b)} \qquad \boldsymbol{\lambda}_0(i) = \boldsymbol{\lambda}_0(i+1) = \cdots = \boldsymbol{\lambda}_0(K) > 0$$

$$\text{(c)} \qquad \boldsymbol{\lambda}_n(i) = \boldsymbol{\lambda}_n(i+1) = \cdots = \boldsymbol{\lambda}_n(K) = 0$$

$$\text{(d)} \qquad \boldsymbol{\lambda}_1(K) = \cdots = \boldsymbol{\lambda}_{n-1}(K) = \boldsymbol{\lambda}_n(K) = 0$$

4.19 CONSTRAINTS

Now let us consider constraint sets $K_u(x^k)$ and $K_v(x^k)$, respectively, given by

$$\begin{aligned} \varphi_\alpha{}^k(x^k, u) &\leqslant 0 \qquad \alpha = 1, 2, \ldots, l \\ \psi_\alpha{}^k(x^k, v) &\leqslant 0 \qquad \alpha = 1, 2, \ldots, m \end{aligned} \tag{4.63}$$

where functions $\varphi_\alpha{}^k$ and $\psi_\alpha{}^k$ are of class C^1 on $G_k \times U$ and $G_k \times V$, respectively.

Suppose that

$$\begin{aligned} \varphi_\alpha{}^k(x^k, p^*(x^k)) &= 0 \qquad \alpha = 1, 2, \ldots, l' \leqslant l \\ \psi_\alpha{}^k(x^k, e^*(x^k)) &= 0 \qquad \alpha = 1, 2, \ldots, m' \leqslant m \end{aligned}$$

at $x^k = \mathbf{x}^*(k)$

We shall suppose also that if $l > r$ and $m > s$, at most r of the $\varphi_\alpha{}^k(x^k, u)$ and s of the $\psi_\alpha{}^k(x^k, v)$ vanish at any point of $G_k \times U$ and $G_k \times V$, respectively, and furthermore that the matrices

$$\left[\frac{\partial \varphi_\alpha{}^k(x^k, u)}{\partial u_\sigma}\right] \qquad \begin{aligned} \alpha &= 1, 2, \ldots, l \\ \sigma &= 1, 2, \ldots, r \end{aligned}$$

$$\left[\frac{\partial \psi_\alpha{}^k(x^k, v)}{\partial v_\sigma}\right] \qquad \begin{aligned} \alpha &= 1, 2, \ldots, m \\ \sigma &= 1, 2, \ldots, s \end{aligned}$$

have maximum rank at $u = p^*(x^k)$, $v = e^*(x^k)$, $x^k = \mathbf{x}^*(k)$.

As a consequence of condition (a) of Theorem 4.1, the Lagrange multiplier rule gives

$$\begin{aligned} \boldsymbol{\lambda}^{\mathrm{T}}(k+1)\frac{\partial \mathbf{f}^k}{\partial u} + \boldsymbol{\mu}^{\mathrm{T}}(k)\frac{\partial \varphi^k}{\partial u} &= 0 \\ \boldsymbol{\lambda}^{\mathrm{T}}(k+1)\frac{\partial \mathbf{f}^k}{\partial v} + \boldsymbol{\nu}^{\mathrm{T}}(k)\frac{\partial \psi^k}{\partial v} &= 0 \end{aligned} \tag{4.64}$$

where $\boldsymbol{\lambda}^{\mathrm{T}}$, $\boldsymbol{\mu}^{\mathrm{T}}$, and $\boldsymbol{\nu}^{\mathrm{T}}$ are row vectors

$$\begin{aligned} &\boldsymbol{\lambda}^{\mathrm{T}} \triangleq (\boldsymbol{\lambda}_0\, \boldsymbol{\lambda}_1 \ldots \boldsymbol{\lambda}_n) \\ &\boldsymbol{\mu}^{\mathrm{T}} \triangleq (\boldsymbol{\mu}_1\, \boldsymbol{\mu}_2 \ldots \boldsymbol{\mu}_{l'}\, 0 \ldots 0) \qquad \text{where } \mu_i \geqslant 0 \\ &\boldsymbol{\nu}^{\mathrm{T}} \triangleq (\boldsymbol{\nu}_1\, \boldsymbol{\nu}_2 \ldots \boldsymbol{\nu}_{m'}\, 0 \ldots 0) \qquad \text{where } \nu_i \leqslant 0 \end{aligned}$$

respectively, and

$$\begin{aligned} \frac{\partial \varphi^k}{\partial u} &\triangleq \left[\frac{\partial \varphi_\alpha{}^k(x^k, u)}{\partial u_\sigma}\right] \qquad \begin{aligned} \alpha &= 1, 2, \ldots, l \\ \sigma &= 1, 2, \ldots, r \end{aligned} \\ \frac{\partial \psi^k}{\partial v} &\triangleq \left[\frac{\partial \psi_\alpha{}^k(x^k, v)}{\partial v_\sigma}\right] \qquad \begin{aligned} \alpha &= 1, 2, \ldots, m \\ \sigma &= 1, 2, \ldots, s \end{aligned} \end{aligned}$$

evaluated at $u = p^*(x^k)$, $v = e^*(x^k)$, $x^k = \mathbf{x}^*(k)$.

Since $\varphi_\alpha{}^k(x^k, p^*(x^k))$, $\alpha = 1, 2, \ldots, l'$, and $\psi_\alpha{}^k(x^k, e^*(x^k))$, $\alpha = 1, 2, \ldots, m'$, have an extremum for $x^k = \mathbf{x}^*(k)$, we may write

$$\begin{aligned} \boldsymbol{\mu}^{\mathrm{T}}(k)\left[\frac{\partial \varphi^k}{\partial \mathbf{x}} + \frac{\partial \varphi^k}{\partial u}\frac{\partial p^*}{\partial \mathbf{x}}\right] &= 0 \\ \boldsymbol{\nu}^{\mathrm{T}}(k)\left[\frac{\partial \psi^k}{\partial \mathbf{x}} + \frac{\partial \psi^k}{\partial v}\frac{\partial e^*}{\partial \mathbf{x}}\right] &= 0 \end{aligned} \tag{4.65}$$

where

$$\begin{aligned} \frac{\partial \varphi^k}{\partial \mathbf{x}} &\triangleq \left[\frac{\partial \varphi_\alpha{}^k(x^k, u)}{\partial x_\sigma}\right] \qquad \begin{aligned} \alpha &= 1, 2, \ldots, l \\ \sigma &= 0, 1, \ldots, n \end{aligned} \\ \frac{\partial \psi^k}{\partial \mathbf{x}} &\triangleq \left[\frac{\partial \psi_\alpha{}^k(x^k, u)}{\partial x_\sigma}\right] \qquad \begin{aligned} \alpha &= 1, 2, \ldots, m \\ \sigma &= 0, 1, \ldots, n \end{aligned} \end{aligned}$$

evaluated at $u = p^*(x^k)$, $v = e^*(x^k)$, $x^k = \mathbf{x}^*(k)$

From (4.64) and (4.65) we obtain

$$\begin{aligned} \boldsymbol{\lambda}^{\mathrm{T}}(k+1)\frac{\partial \mathbf{f}^k}{\partial u}\frac{\partial p^*}{\partial \mathbf{x}} &= \boldsymbol{\mu}^{\mathrm{T}}(k)\frac{\partial \varphi^k}{\partial \mathbf{x}} \\ \boldsymbol{\lambda}^{\mathrm{T}}(k+1)\frac{\partial \mathbf{f}^k}{\partial v}\frac{\partial e^*}{\partial \mathbf{x}} &= \boldsymbol{\nu}^{\mathrm{T}}(k)\frac{\partial \psi^k}{\partial \mathbf{x}} \end{aligned}$$

or equivalently

$$\left(\frac{\partial \mathbf{f}^k}{\partial u}\frac{\partial p^*}{\partial \mathbf{x}}\right)^{\mathrm{T}} \boldsymbol{\lambda}(k+1) = \left(\frac{\partial \varphi^k}{\partial \mathbf{x}}\right)^{\mathrm{T}} \boldsymbol{\mu}(k)$$
$$\left(\frac{\partial \mathbf{f}^k}{\partial v}\frac{\partial e^*}{\partial \mathbf{x}}\right)^{\mathrm{T}} \boldsymbol{\lambda}(k+1) = \left(\frac{\partial \psi^k}{\partial \mathbf{x}}\right)^{\mathrm{T}} \boldsymbol{\nu}(k) \tag{4.66}$$

where $\boldsymbol{\mu}$ and $\boldsymbol{\nu}$ are column vectors

$$\boldsymbol{\mu} = \begin{pmatrix} \mu_1 \\ \mu_2 \\ \cdot \\ \cdot \\ \cdot \\ \mu_{l'} \\ 0 \\ \cdot \\ \cdot \\ \cdot \\ 0 \end{pmatrix} \qquad \boldsymbol{\nu} = \begin{pmatrix} \nu_1 \\ \nu_2 \\ \cdot \\ \cdot \\ \cdot \\ \nu_{m'} \\ 0 \\ \cdot \\ \cdot \\ \cdot \\ 0 \end{pmatrix}$$

Finally, by (4.66) adjoint equation (4.51) may be written as

$$\boldsymbol{\lambda}(k+1) - \boldsymbol{\lambda}(k) = -\left(\frac{\partial \mathbf{f}^k}{\partial \mathbf{x}}\right)^{\mathrm{T}} \boldsymbol{\lambda}(k+1) - \left(\frac{\partial \varphi^k}{\partial \mathbf{x}}\right)^{\mathrm{T}} \boldsymbol{\mu}(k) - \left(\frac{\partial \psi^k}{\partial \mathbf{x}}\right)^{\mathrm{T}} \boldsymbol{\nu}(k)$$
$$k = i, i+1, \ldots, K-1 \tag{4.67}$$

If the constraints are independent of state x, the adjoint equation reduces to

$$\boldsymbol{\lambda}(k+1) - \boldsymbol{\lambda}(k) = -\left(\frac{\partial \mathbf{f}^k}{\partial \mathbf{x}}\right)^{\mathrm{T}} \boldsymbol{\lambda}(k+1) \qquad k = i, i+1, \ldots, K-1 \tag{4.68}$$

4.20 EXAMPLE 4.1

Let us consider a multistage game governed by

$$\begin{aligned} \mathbf{x}_1(k+1) - \mathbf{x}_1(k) &= \mathbf{x}_2(k) \\ \mathbf{x}_2(k+1) - \mathbf{x}_2(k) &= \mathbf{u}(k) + \mathbf{v}(k) \\ \mathbf{x}_3(k+1) - \mathbf{x}_3(k) &= 1 \end{aligned} \tag{4.69}$$

Let the cost be given by

$$\mathbf{x}_0(k+1) - \mathbf{x}_0(k) = \mathbf{x}_1(k)\mathbf{x}_2(k) + \mathbf{u}^2(k) - \mathbf{v}^2(k) \tag{4.70}$$

and the control constraints by

$$K_u = K_v = E^1$$

A path in augmented state-space E^4, generated by a strategy pair that satisfies the necessary conditions for optimality, will be represented by $\mathbf{x}^*$: $k \to x^k = \mathbf{x}^*(k)$, $k \in \{i, i+1, \ldots, K\}$, where $\mathbf{x}^* \triangleq (\mathbf{x}_0^*, \mathbf{x}_1^*, \mathbf{x}_2^*, \mathbf{x}_3^*)$, and the corresponding controls will be denoted by $\mathbf{u}^*$: $k \to u = \mathbf{u}^*(k)$, $\mathbf{v}^*$: $k \to v = \mathbf{v}^*(k)$.

In view of condition (c) of Theorem 4.1, the $\mathscr{H}$-function is

$$\begin{aligned}\mathscr{H}^{k+1} = {} & \mathbf{x}_1(k)\mathbf{x}_2(k) + \mathbf{u}^2(k) - \mathbf{v}^2(k) + \boldsymbol{\lambda}_1(k+1)\mathbf{x}_2(k) \\ & + \boldsymbol{\lambda}_2(k+1)(\mathbf{u}(k) + \mathbf{v}(k))\end{aligned} \tag{4.71}$$

and the adjoint equations for $\boldsymbol{\lambda}_1$ and $\boldsymbol{\lambda}_2$ are

$$\begin{aligned}\boldsymbol{\lambda}_1(k+1) - \boldsymbol{\lambda}_1(k) &= -\mathbf{x}_2^*(k) \\ \boldsymbol{\lambda}_2(k+1) - \boldsymbol{\lambda}_2(k) &= -\mathbf{x}_1^*(k) - \boldsymbol{\lambda}_1(k+1)\end{aligned} \tag{4.72}$$

for $k = i, i+1, \ldots, K-1$.

From condition (a) of Theorem 4.1 we obtain

$$\begin{aligned}2\mathbf{u}^*(k) + \boldsymbol{\lambda}_2(k+1) &= 0 \\ -2\mathbf{v}^*(k) + \boldsymbol{\lambda}_2(k+1) &= 0\end{aligned} \tag{4.73}$$

which implies that

$$\mathbf{u}^*(k) = -\mathbf{v}^*(k) \tag{4.74}$$

Condition (d) of Theorem 4.1 yields

$$\boldsymbol{\lambda}_1(K) = \boldsymbol{\lambda}_2(K) = 0 \tag{4.75}$$

Let

$$\begin{aligned}\mathbf{x}_0^*(K) &= x_0^K = C \\ \mathbf{x}_1^*(K) &= x_1^K \\ \mathbf{x}_2^*(K) &= x_2^K\end{aligned} \tag{4.76}$$

Now we deduce from (4.69)–(4.75) that

at stage $K-1$

$$\begin{aligned}\mathbf{x}_2^*(K-1) &= \mathbf{x}_2^*(K) = x_2^K \\ \mathbf{x}_1^*(K-1) &= \mathbf{x}_1^*(K) - \mathbf{x}_2^*(K-1) = x_1^K - x_2^K \\ \mathbf{x}_0^*(K-1) &= C - x_2^K(x_1^K - x_2^K)\end{aligned}$$

and

$$\boldsymbol{\lambda}_1(K-1) = x_2^K$$
$$\boldsymbol{\lambda}_2(K-1) = x_1^K - x_2^K$$

at stage $K-2$

$$\mathbf{x}_2^*(K-2) = \mathbf{x}_2^*(K-1) = x_2^K$$
$$\mathbf{x}_1^*(K-2) = \mathbf{x}_1^*(K-1) - x_2^K = x_1^K - 2x_2^K$$
$$\mathbf{x}_0^*(K-2) = -x_2^K(x_1^K - 2x_2^K) + \mathbf{x}_0^*(K-1)$$
$$= C - x_2^K(2x_1^K - 3x_2^K)$$

and

$$\boldsymbol{\lambda}_1(K-2) = x_2^K + x_2^K = 2x_2^K$$
$$\boldsymbol{\lambda}_2(K-2) = x_1^K - 2x_2^K + x_2^K + x_1^K - x_2^K = 2(x_1^K - x_2^K)$$

More generally, one can prove readily by recursive arguments that

at stage $K-m \quad (0 \leqslant m \leqslant K-i)$

$$\mathbf{x}_2^*(K-m) = x_2^K$$
$$\mathbf{x}_1^*(K-m) = x_1^K - mx_2^K \tag{4.77}$$
$$\mathbf{x}_0^*(K-m) = C - mx_2^K\{x_1^K - \tfrac{1}{2}(m+1)x_2^K\}$$

and

$$\boldsymbol{\lambda}_1(K-m) = mx_2^K$$
$$\boldsymbol{\lambda}_2(K-m) = m(x_1^K - x_2^K) \tag{4.78}$$

Hence, from (4.73) and the second of Eqs. (4.78), we have

$$\mathbf{u}^*(K-m) = -\mathbf{v}^*(K-m) = -[\tfrac{1}{2}(m-1)](x_1^K - x_2^K) \qquad m = 1, 2, \ldots, K-i \tag{4.79}$$

The Value of the game for play starting at stage k is

$$V^*(\mathbf{x}^*(k)) = \mathbf{x}_0^*(K) - \mathbf{x}_0^*(k) = (K-k)x_2^K\{x_1^K - \tfrac{1}{2}(K-k+1)x_2^K\}$$

Now let

$$\mathbf{x}_0^*(k) = x_0^k, \qquad \mathbf{x}_1^*(k) = x_1^k, \qquad \mathbf{x}_2^*(k) = x_2^k$$

From (4.77) we deduce

$$x_2^k = x_2^K, \qquad x_1^k = x_1^K - (K-k)x_2^K$$

and hence, the Value of the game for play starting at stage k may be written as

$$V^*(\mathbf{x}^*(k)) = (K-k)x_2^k\{x_1^k + \tfrac{1}{2}(K-k-1)x_2^k\} \tag{4.80}$$

It follows that the equation of surface $\Sigma_k(C)$ is

$$x_0^k + (K-k)x_2^k\{x_1^k + \tfrac{1}{2}(K-k-1)x_2^k\} = C \qquad (4.81)$$
$$k = 0, 1, 2, \ldots, K$$

Surface $\Sigma^k(C)$ is shown on Figs. 4.7–4.9 for $k \neq K, K-1$, $k = K-1$, and $k = K$, respectively, in x_0–x_1–x_2 space.

The sets $\Omega_P(K-m)$ and $\Omega_E(K-m)$, $m = 0, 1, \ldots, K-i-1$ are defined by

$$\Omega_P(K-m) = \{\mathbf{x}(K-m){:}\ \ \mathbf{x}(K-m) = \mathbf{x}^*(K-m-1) + \mathbf{f}^{K-m-1}(\mathbf{x}^*(K-m-1), \mathbf{u}^*(K-m-1), \mathbf{v}(K-m-1)) \ \forall \mathbf{v}(K-m-1) \in K_v\}$$

$$\Omega_E(K-m) = \{\mathbf{x}(K-m){:}\ \ \mathbf{x}(K-m) = \mathbf{x}^*(K-m-1) + \mathbf{f}^{K-m-1}(\mathbf{x}^*(K-m-1), \mathbf{u}(K-m-1), \mathbf{v}^*(K-m-1)) \ \forall \mathbf{u}(K-m-1) \in K_u\},$$

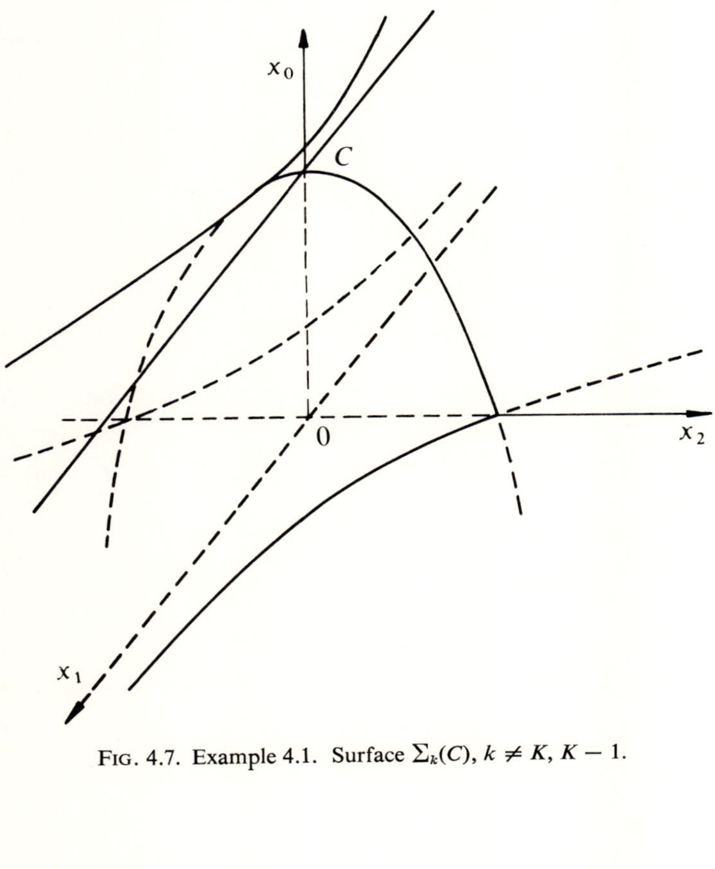

FIG. 4.7. Example 4.1. Surface $\Sigma_k(C)$, $k \neq K, K-1$.

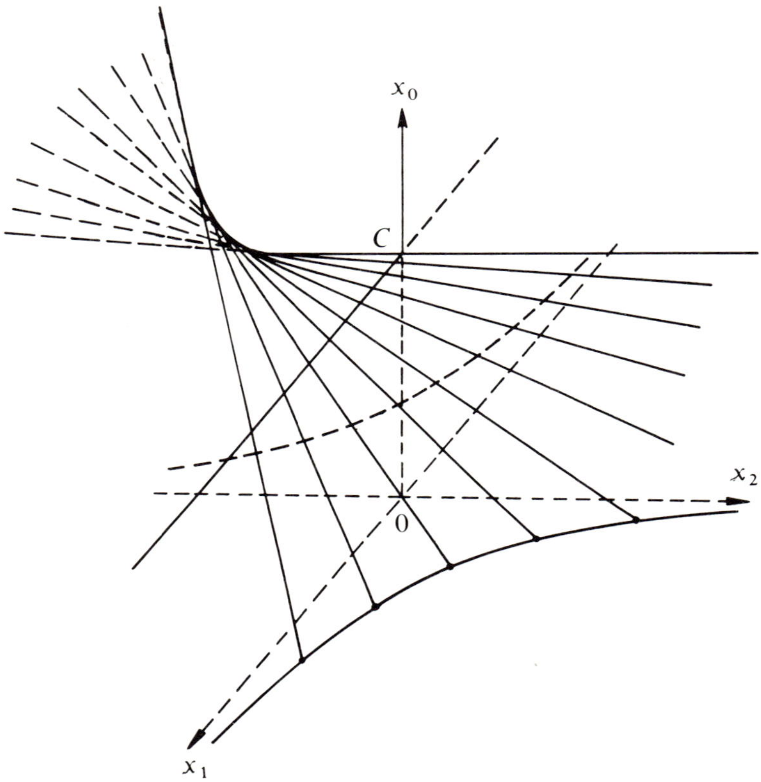

FIG. 4.8. Example 4.1. Surface $\Sigma_{K-1}(C)$.

For instance, let us consider $\Omega_P(K - m)$. In view of Eqs. (4.69) and (4.70), $\Omega_P(K - m)$ is the set of points

$$\mathbf{x}(K - m) = (\mathbf{x}_0(K - m), \mathbf{x}_1(K - m), \mathbf{x}_2(K - m), K - m)$$

such that

$$\mathbf{x}_0(K - m) = \mathbf{x}_0{}^*(K - m - 1) + \mathbf{x}_1{}^*(K - m - 1)\mathbf{x}_2{}^*(K - m - 1) + \mathbf{u}^{*2}(K - m - 1) - \mathbf{v}^2(K - m - 1) \quad (4.82)$$

$$\mathbf{x}_1(K - m) = \mathbf{x}_1{}^*(K - m - 1) + \mathbf{x}_2{}^*(K - m - 1) \quad (4.83)$$

$$\mathbf{x}_2(K - m) = \mathbf{x}_2{}^*(K - m - 1) + \mathbf{u}^*(K - m - 1) + \mathbf{v}(K - m - 1) \quad (4.84)$$

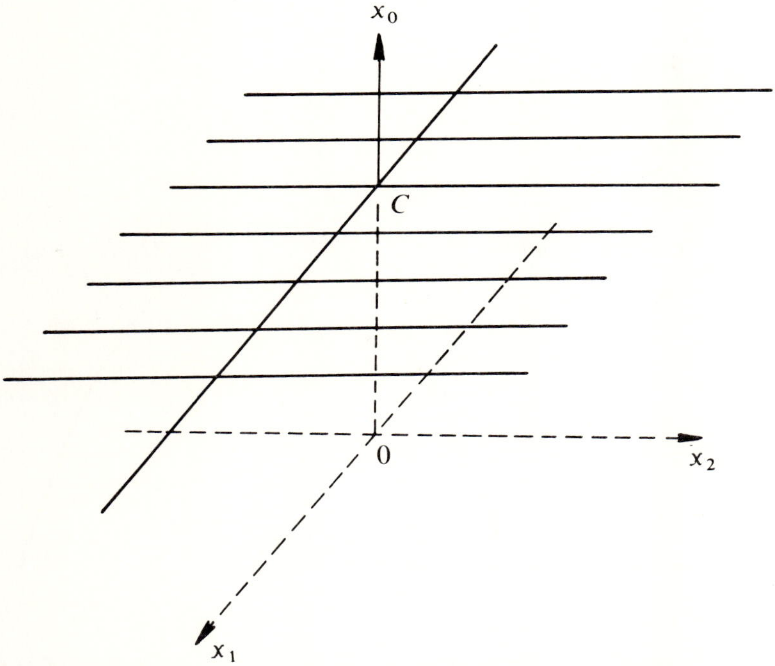

FIG. 4.9. Example 4.1. Surface $\Sigma_K(C) \cdot \Sigma_K(C)$ is a plane perpendicular to the x_0 axis.

Equations (4.82)–(4.84) together with (4.77) and (4.79) imply that

$$\begin{aligned} \mathbf{x}_0(K - m) = C &- (m + 1)x_2{}^K\{x_1{}^K - \tfrac{1}{2}(m + 2)x_2{}^K\} \\ &+ [x_1{}^K - (m + 1)x_2{}^K]x_2{}^K \\ &+ \tfrac{1}{4}m^2(x_1{}^K - x_2{}^K)^2 - \mathbf{v}^2(K - m - 1) \end{aligned} \tag{4.85}$$

$$\mathbf{x}_1(K - m) = \mathbf{x}_1{}^*(K - m) = x_1{}^K - mx_2{}^K \tag{4.86}$$

$$\mathbf{x}_2(K - m) = x_2{}^K - \tfrac{1}{2}m(x_1{}^K - x_2{}^K) + \mathbf{v}(K - m - 1) \tag{4.87}$$

or equivalently

$$\begin{aligned} \mathbf{x}_0(K - m) &= \mathbf{x}_0{}^*(K - m) \\ &+ [\tfrac{1}{2}m(x_1{}^K - x_2{}^K) - \mathbf{v}(K - m - 1)][\tfrac{1}{2}m(x_1{}^K - x_2{}^K) + \mathbf{v}(K - m - 1)] \end{aligned} \tag{4.88}$$

$$\mathbf{x}_1(K - m) = \mathbf{x}_1{}^*(K - m) \tag{4.89}$$

$$\mathbf{x}_2(K - m) = \mathbf{x}_2{}^*(K - m) - \tfrac{1}{2}m(x_1{}^K - x_2{}^K) + \mathbf{v}(K - m - 1) \tag{4.90}$$

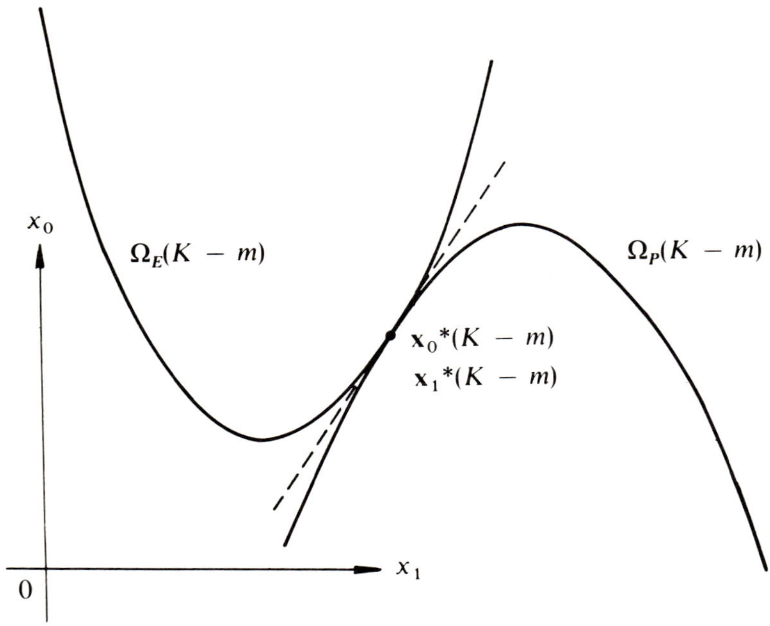

FIG. 4.10. Example 4.1. Intersections of the sets $\Omega_P(K-m)$ and $\Omega_E(K-m)$ with the x_0–x_1 plane.

From (4.88)–(4.90), by eliminating $\mathbf{v}(K-m-1)$ between (4.88) and (4.90), we obtain the equations defining $\Omega_P(K-m)$, namely

$$\mathbf{x}_0(K-m) - \mathbf{x}_0^*(K-m) = [\mathbf{x}_2^*(K-m) - \mathbf{x}_2(K-m) - m(x_1^K - x_2^K)] \times [\mathbf{x}_2(K-m) - \mathbf{x}_2^*(K-m)] \quad (4.91)$$

$$\mathbf{x}_1(K-m) = \mathbf{x}_1^*(K-m) \quad (4.92)$$

Similarly, one can see that $\Omega_E(K-m)$ is defined by

$$\mathbf{x}_0(K-m) - \mathbf{x}_0^*(K-m) = [\mathbf{x}_2(K-m) - \mathbf{x}_2^*(K-m) - m(x_1^K - x_2^K)] \times [\mathbf{x}_2(K-m) - \mathbf{x}_2^*(K-m)] \quad (4.93)$$

$$\mathbf{x}_1(K-m) = \mathbf{x}_1^*(K-m) \quad (4.94)$$

These sets are parabolic cylinders parallel to the x_2-axis. Their intersections with the x_0–x_1 plane are sketched on Fig. 4.10. It is readily seen that Assumption 4.2 is satisfied.

V *Some Geometric Aspects of Qualitative Games*

5.1 MAP OF A GAME

Given any strategy $r_E \in R_E$, let us associate with it a set of points $A(r_E)$ in G; namely,

$$A(r_E) \triangleq \{x^i: \quad \exists r_P \in R_P, \qquad (r_P, r_E) \in \mathscr{T}_P(x^i)\}$$

Similarly, with every strategy $r_P \in R_P$, let us associate a set $B(r_P)$; namely,

$$B(r_P) \triangleq \{x^i: \quad \exists r_E \in R_E, \qquad (r_P, r_E) \in \mathscr{T}_E(x^i)\}$$

Then let us define the following sets:

$$A \triangleq \bigcap_{r_E \in R_E} A(r_E), \qquad B \triangleq \bigcap_{r_P \in R_P} B(r_P)$$

Clearly, we have

$$A = \{x^i: \quad \forall r_E \in R_E \quad \exists r_P \in R_P, \quad (r_P, r_E) \in \mathscr{T}_P(x^i)\}$$
$$B = \{x^i: \quad \forall r_P \in R_P \quad \exists r_E \in R_E, \quad (r_P, r_E) \in \mathscr{T}_E(x^i)\}$$

It follows that†

$$\text{comp } A = \{x^i: \quad \exists r_E \in R_E \quad \forall r_P \in R_P, \quad (r_P, r_E) \notin \mathscr{T}_P(x^i)\}$$
$$\text{comp } B = \{x^i: \quad \exists r_P \in R_P \quad \forall r_E \in R_E, \quad (r_P, r_E) \notin \mathscr{T}_E(x^i)\}$$

Definition 5.1. Sets A, B, comp A, comp B, and the intersections $A \cap B$ and comp $A \cap$ comp B, constitute the *map of the game*.

† Hereafter comp S will mean the complement of set S in G. Also, the boundary ∂S of a set S will be defined by $\partial S \triangleq \bar{S} \cap \overline{\text{comp } S}$, where $\bar{S}$ and $\overline{\text{comp } S}$ are the closures of S and comp S, respectively, in the topology induced by E^n on G (see Appendix). Note that $\partial S \subset G$.

5.2 OPTIMAL STRATEGIES IN A QUALITATIVE GAME, SETS OF THE GAME

Definition 5.2. In a qualitative game, we shall say that $r_P{}^*(x^i) \in R_P$ is an *optimal strategy for player J_P at point x^i*, if $(r_P{}^*(x^i), r_E)$ is strongly playable for J_P at that point, no matter what $r_E \in R_E$; that is

$$(r_P{}^*(x^i), r_E) \in \mathscr{T}_P{}^*(x^i) \qquad \forall r_E \in R_E \tag{5.1}$$

Likewise we shall say that $r_E{}^*(x^i) \in R_E$ is an *optimal strategy for player J_P at point x^i*, if $(r_P, r_E{}^*(x^i))$ is strongly playable for J_E at that point, no matter what $r_P \in R_P$; that is

$$(r_P, r_E{}^*(x^i)) \in \mathscr{T}_E{}^*(x^i) \qquad \forall r_P \in R_P \tag{5.2}$$

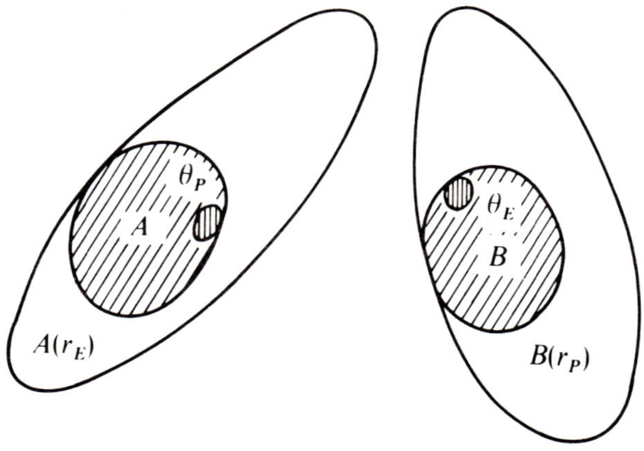

FIG. 5.1. Map of a game.

Recall that strong playability for J_P (or J_E) at x^i implies weak playability for J_P (or J_E) at that point. Moreover, the concepts of strongly and weakly playable pair (r_P, r_E) for J_E (or J_P) at x^i coincide if $\theta_E = \varnothing$ (or $\theta_P = \varnothing$), or if $\theta_P \cap \theta_E = \varnothing$ and (r_P, r_E) generates a unique describing curve from x^i.

Returning to the definition of the map of a game and to the definition of a playable strategy pair, we see that (5.1) implies that

$$x^i \in A \cap \text{comp } B \tag{5.3}$$

This follows from the fact that

$$(r_P^*(x^i), r_E) \in \mathscr{T}_P^*(x^i) \Rightarrow (r_P^*(x^i), r_E) \notin \mathscr{T}_E(x^i)$$

and hence (5.1) implies that $x^i \in \text{comp } B$.

On the other hand, since

$$(r_P^*(x^i), r_E) \in \mathscr{T}_P^*(x^i) \Rightarrow (r_P^*(x^i), r_E) \in \mathscr{T}_P(x^i)$$

and since $(r_P^*(x^i), r_E) \in \mathscr{T}_P(x^i)$ $\forall r_E \in R_E$ implies that $\forall r_E \in R_E$ $\exists r_P \in R_P$ [namely, $r_P = r_P^*(x^i)$] such that $(r_P, r_E) \in \mathscr{T}_P(x^i)$, it follows that $x^i \in A$.

By similar arguments one can prove that (5.2) implies that

$$x^i \in B \cap \text{comp } A \tag{5.4}$$

Definition 5.3.

$$S_P \triangleq \{x^i\colon \ \exists r_P^*(x^i)\}$$
$$S_E \triangleq \{x^i\colon \ \exists r_E^*(x^i)\}$$

Sets S_P and S_E will be called the *sets of the game*.

Let us note first that $(\theta_P \cap \theta_E) \cap (S_P \cup S_E) = \varnothing$. If $\theta_P \cap \theta_E = \varnothing$ and $\theta_P \neq \varnothing$ then $\theta_P \subseteq S_P$; and if $\theta_P \cap \theta_E = \varnothing$ and $\theta_E \neq \varnothing$ then $\theta_E \subseteq S_E$.

If target θ_P (or θ_E) is empty, the corresponding sets S_P (or S_E) may be empty or not.

Furthermore, it follows from the remarks of Sec. 5.2 that

$$S_P \subseteq A \cap \text{comp } B \tag{5.5}$$

and

$$S_E \subseteq B \cap \text{comp } A \tag{5.6}$$

Hence,

$$S_P \cap S_E = \varnothing \tag{5.7}$$

Definition 5.4. Let there be given a fixed set $X_P \subseteq S_P$. We shall say that $\rho_P^* \in R_P$ is *optimal on* X_P, if

$$\forall x^i \in X_P \quad \forall r_E \in R_E, \qquad (\rho_P^*, r_E) \in \mathscr{T}_P^*(x^i) \tag{5.8}$$

Likewise, let there be given a fixed set $X_E \subseteq S_E$. We shall say that $\rho_E^* \in R_E$ is *optimal on* X_E, if

$$\forall x^i \in X_E \quad \forall r_P \in R_P, \qquad (r_P, \rho_E^*) \in \mathscr{T}_E^*(x^i) \tag{5.9}$$

In other words a strategy for player J_P that is optimal on X_P is optimal at all points $x^i \in X_P$; that is, $\exists r_P^*(x^i)$ such that $r_P^*(x^i) = \rho_P^*$ for all $x^i \in X_P$. Likewise, a strategy for player J_E that is optimal on X_E is optimal at all points $x^i \in X_E$; that is, $\exists r_E^*(x^i)$ such that $r_E^*(x^i) = \rho_E^*$ for all $x^i \in X_E$.

5.3 SOME PROPERTIES OF SETS OF THE GAME

Definition 5.5. A $\mathscr{P}$*-path* is a path that emanates from $x^i \in S_P$ and is generated by $(r_P^*(x^i), r_E)$, $r_E \in R_E$; similarly, an $\mathscr{E}$*-path* is a path that

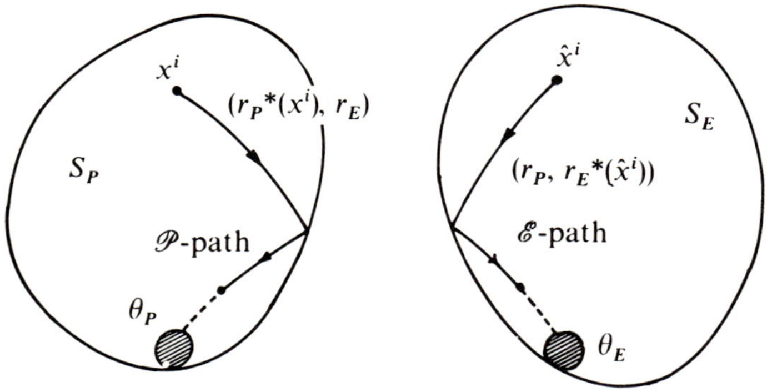

FIG. 5.2. Sets of the game, $\mathscr{P}$ and $\mathscr{E}$-paths.

emanates from $\hat{x}^i \in S_E$ and is generated by $(r_P, r_E^*(\hat{x}^i))$, $r_P \in R_P$. $\mathscr{P}$ and $\mathscr{E}$-paths that reach targets θ_P and θ_E, respectively, will be called $\mathscr{P}$*-optimal* and $\mathscr{E}$*-optimal* paths, respectively.

Now, let us prove

Lemma 5.1. *No point of a* $\mathscr{P}$*-path belongs to comp* S_P;

Lemma 5.2. *No point of an* $\mathscr{E}$*-path belongs to comp* S_E.

Consider path $\pi_{\mathscr{P}}^{is}$, emanating from $x^i \in S_P$, generated by strategy pair $(r_P^*(x^i), r_E)$, and ending at point x^s; and suppose that there exists a point x^j in $\pi_{\mathscr{P}}^{is} \cap \text{comp } S_P$.

Let $\pi_{\mathscr{P}}^{ij} \subset \pi_{\mathscr{P}}^{is}$ be the path that emanates from x^i, ends at x^j, and is given by the same function that defines $\pi_{\mathscr{P}}^{is}$ between x^i and x^j.

It follows from the definition of S_P that

$$\text{comp } S_P = \{x\colon \quad \forall r_P \quad \exists r_E, \quad (r_P, r_E) \notin \mathscr{T}_P^*(x)\}$$

and hence there exists $\hat{r}_E \in R_E$ such that $(r_P^*(x^i), \hat{r}_E) \notin \mathscr{T}_P^*(x^j)$

Then it follows from the definition of a strongly playable pair for J_P at x^j, that either

(i) $\exists \hat{\mathbf{x}} \in \gamma(x^j, r_P^*(x^i), \hat{r}_E), \qquad \Gamma_{\hat{\mathbf{x}}}(x^j, r_P^*(x^i), \hat{r}_E) \cap \theta_P = \varnothing$

or

(ii) $\forall \mathbf{x} \in \gamma(x^j, r_P^*(x^i), \hat{r}_E), \qquad \Gamma_{\hat{\mathbf{x}}}(x^j, r_P^*(x^i), \hat{r}_E) \cap \theta_P \neq \varnothing$
and there exists $\hat{\mathbf{x}} \in \gamma(x^j, r_P^*(x^i), \hat{r}_E)$ that generates from x^j a terminating path whose end state belongs to θ_E.

In cases (i) and (ii), it follows from Assumption 1.1 that there exist $\rho_E \in R_E$ and a function $\xi \in \gamma(x^i, r_P^*(x^i), \rho_E)$ such that

$$\Gamma_{\xi}(x^i, r_P^*(x^i), \rho_E) = \pi_{\mathscr{P}}^{ij} \cup \Gamma_{\hat{\mathbf{x}}}(x^j, r_P^*(x^i), \hat{r}_E)$$

Consequently, *in Case (i)*

$$\Gamma_{\xi}(x^i, r_P^*(x^i), \rho_E) \cap \theta_P = \varnothing$$

and *in Case (ii)* ξ generates from x^i a terminating path whose end state belongs to θ_E.

Both cases contradict the fact that $r_P^*(x^i)$ is an optimal strategy for J_P at point x^i. Hence Lemma 5.1 is established.

Lemma 5.2 can be proved by invoking similar arguments.

5.4 SURFACE OF THE GAME

Definition 5.6. In a qualitative game, the set Ξ, defined by

$$\Xi \triangleq \partial S_P \cap \partial S_E \tag{5.10}$$

will be termed the *surface of the game*.

Recall that

$$\partial S_P \triangleq \bar{S}_P \cap \overline{\text{comp } S_P}$$

$$\partial S_E \triangleq \bar{S}_E \cap \overline{\text{comp } S_E}$$

where comp S, $S = S_P$ or $S = S_E$, means the complement of S in G. Of course, Ξ may be empty.

One can readily establish

Lemma 5.3. *Ξ is the intersection of the closures of S_P and S_E; namely,*

$$\Xi = \bar{S}_P \cap \bar{S}_E \tag{5.11}$$

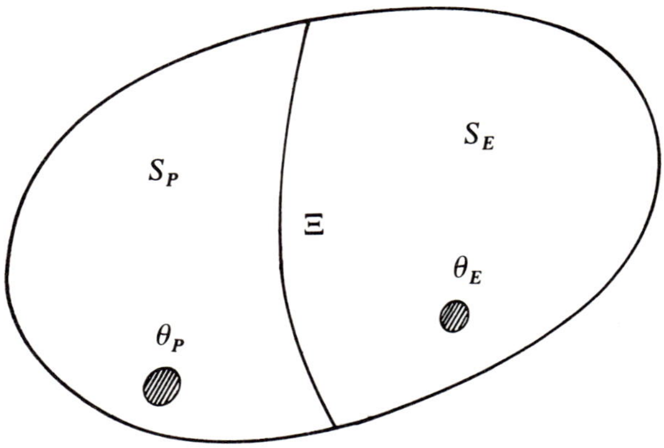

FIG. 5.3. Game surface Ξ.

From (5.10) we have

$$\begin{aligned} \Xi &= (\bar{S}_P \cap \overline{\text{comp } S_P}) \cap (\bar{S}_E \cap \overline{\text{comp } S_E}) \\ &= \bar{S}_P \cap \bar{S}_E \cap \overline{\text{comp } S_P} \cap \overline{\text{comp } S_E} \end{aligned} \tag{5.12}$$

Since $S_P \cap S_E = \varnothing$, it follows that

$$S_P \subseteq \text{comp } S_E$$
$$S_E \subseteq \text{comp } S_P$$

Hence,

$$\bar{S}_P \subseteq \overline{\text{comp } S_E}$$
$$\bar{S}_E \subseteq \overline{\text{comp } S_P}$$

and accordingly,

$$\bar{S}_P \cap \overline{\text{comp } S_E} = \bar{S}_P$$
$$\bar{S}_E \cap \overline{\text{comp } S_P} = \bar{S}_E$$

Finally, by substituting in (5.12), we obtain (5.11) which proves Lemma 5.3.

Another property of Ξ is embodied in

Lemma 5.4. *Ξ belongs to the intersection of the boundaries of A and B; namely,*

$$\Xi \subseteq \partial A \cap \partial B \tag{5.13}$$

From (5.5) and (5.6) we have

$$S_P \subseteq A, \qquad S_P \subseteq \text{comp } B$$
$$S_E \subseteq B, \qquad S_E \subseteq \text{comp } A$$

and hence,

$$\bar{S}_P \subseteq \bar{A}, \qquad \bar{S}_P \subseteq \overline{\text{comp } B}$$
$$\bar{S}_E \subseteq \bar{B}, \qquad \bar{S}_E \subseteq \overline{\text{comp } A}$$

Then

$$\bar{S}_P \cap \bar{S}_E \subseteq \bar{A} \cap \overline{\text{comp } A} \triangleq \partial A \tag{5.14}$$

$$\bar{S}_P \cap \bar{S}_E \subseteq \bar{B} \cap \overline{\text{comp } B} \triangleq \partial B \tag{5.15}$$

Relations (5.14) and (5.15), together with (5.11), establish Lemma 5.4.

We have also

Lemma 5.5. *If $\bar{S}_P = \overline{\text{comp } S_E}$ and $\bar{S}_E = \overline{\text{comp } S_P}$, then*

$$\Xi = \partial A = \partial B \tag{5.16}$$

From (5.5) and (5.6) one can readily deduce that

$$\text{comp } A \subseteq \text{comp } S_P, \qquad B \subseteq \text{comp } S_P$$
$$\text{comp } B \subseteq \text{comp } S_E, \qquad A \subseteq \text{comp } S_E$$

Hence,

$$\overline{\text{comp } A} \subseteq \overline{\text{comp } S_P}, \qquad \bar{B} \subseteq \overline{\text{comp } S_P}$$
$$\overline{\text{comp } B} \subseteq \overline{\text{comp } S_E}, \qquad \bar{A} \subseteq \overline{\text{comp } S_E}$$

It follows that

$$\partial A \triangleq \bar{A} \cap \overline{\text{comp } A} \subseteq \overline{\text{comp } S_E} \cap \overline{\text{comp } S_P} \tag{5.17}$$

$$\partial B \triangleq \bar{B} \cap \overline{\text{comp } B} \subseteq \overline{\text{comp } S_P} \cap \overline{\text{comp } S_E} \tag{5.18}$$

By substituting (5.17) and (5.18) in (5.14) and (5.15), respectively, we obtain

$$\bar{S}_P \cap \bar{S}_E \subseteq \partial A \subseteq \overline{\text{comp } S_P} \cap \overline{\text{comp } S_E}$$

$$\bar{S}_P \cap \bar{S}_E \subseteq \partial B \subseteq \overline{\text{comp } S_P} \cap \overline{\text{comp } S_E}$$

Finally, if $\bar{S}_P = \overline{\text{comp } S_E}$ and $\bar{S}_E = \overline{\text{comp } S_P}$, we have $\bar{S}_P \cap \bar{S}_E = \partial A = \partial B = \overline{\text{comp } S_P} \cap \overline{\text{comp } S_E}$ and so

$$\Xi = \bar{S}_P \cap \bar{S}_E = \partial A = \partial B.$$

Of course, if $S_P = \text{comp } S_E$, the assumptions of Lemma 5.5 are satisfied and $\Xi = \partial A = \partial B$.

5.5 A SIMILARITY BETWEEN QUALITATIVE AND QUANTITATIVE GAMES

It is readily seen that Lemmas 5.1 and 5.2 exhibit strong similarities to Corollaries 2.2 and 2.3, respectively, provided we invoke the correspondence between

$$S_P \quad \text{and} \quad B/\Sigma(C)$$

$$S_E \quad \text{and} \quad A/\Sigma(C)$$

Also, in differential qualitative games, under certain conditions, properties of Ξ will appear to be similar to those of a Σ-surface.

However, there are differences between these corresponding sets. Some of them are listed below:

(i) Ξ may be empty, whereas $\Sigma(C)$ cannot be empty.

(ii) Ξ is unique, but there is a one-parameter family of $\Sigma(C)$.

(iii) In one target qualitative games,† S_P or S_E may be empty, whereas $B/\Sigma(C)$ and $A/\Sigma(C)$ are never empty.

(iv) S_P and S_E are unique, whereas there is a one-parameter family of $B/\Sigma(C)$ and $A/\Sigma(C)$.

These differences play a role in some of the derivations of the next chapter.

Many of the results which have been established for regular points of a Σ-surface will be seen to apply for regular points of Ξ, provided we replace $\Sigma(C)$ by Ξ, $A/\Sigma(C)$ by S_E, and $B/\Sigma(C)$ by S_P.

† Either $\theta_E = \varnothing$ or $\theta_P = \varnothing$.

5.6 PROBLEM STATEMENT

Now we are ready to state the problem of qualitative games. We shall consider separately two target and one target games, that is Case (a) and Case (b). Let us first list the different situations that players J_P and J_E can meet in these two cases.

Case (*a*) $\theta_P \neq \varnothing$ and $\theta_E \neq \varnothing$

a.1 If $x^i \in S_P$ and if J_P plays optimally, then J_P is sure to win, no matter what the strategy of J_E.

a.2 If $x^i \in S_E$ and if J_E plays optimally, then J_E is sure to win, no matter what the strategy of J_P.

a.3 If $x^i \notin S_P \cup S_E$, then there is no optimal strategy at that point for J_P and for J_E.

Case (*b*) with $\theta_P \neq \varnothing$ and $\theta_E = \varnothing$

b.1 If $x^i \in S_P$ and if J_P plays optimally, then J_P is sure to win, no matter what the strategy of J_E.

b.2 If $x^i \in S_E$ and if J_E plays optimally, then J_E can prevent J_P from winning.

b.3 If $x^i \notin S_P \cup S_E$, then there is no optimal strategy at that point for J_P and for J_E.

Recall that $S_P \cap S_E = \varnothing$ in Cases (a) and (b). Hence, Cases a.1 and a.2 on one hand, and b.1 and b.2 on the other hand, exclude each other.

In Cases a.1, a.2, b.1, and b.2, there is an optimal strategy defined for J_P or J_E at the initial state x^i.

In Cases a.1, a.2, and b.1, a winner exists, and this can be assured at the outset of play.

Hence, at the beginning of play the following questions arise:

1. Does $x^i \in S_P \cup S_E$ or does $x^i \notin S_P \cup S_E$?
2. If $x^i \in S_P \cup S_E$, does $x^i \in S_P$ or does $x^i \in S_E$? That is, is there potentially a winner, and who is he?
3. If $x^i \in S_P$, that is if J_P is potentially the winner, what strategy is an optimal one for him? And if $x^i \in S_E$, what strategy is an optimal one for J_E? That is, J_P in one case, and J_E in the other case, must be able to determine an optimal strategy for himself.

In general these questions are difficult to answer. However, questions 1 and 2 are simplified if the assumptions of Lemma 5.5 are satisfied, that is if $\bar{S}_P = \overline{\text{comp } S_E}$ and $\bar{S}_E = \overline{\text{comp } S_P}$. Then S_P and S_E have the same boundary, and the first problem reduces to one of determining this boundary, namely the game surface Ξ.

Accordingly, in the next chapter, we shall focus our attention on further properties of Ξ which hold for differential games, with a view to constructing Ξ. Towards that end the remarks of Section 5.5 will be useful.

VI Differential Qualitative Games

6.1 RULES AND OBJECTIVES OF THE GAME

As in Chapter 3, we shall call $x = (x_1, x_2, \ldots, x_n)$ the state vector of the game in n-dimensional Euclidean space E^n, where $x_n = t$ will denote the time variable. Let there be given the equation

$$dx/d\tau = f(x, u, v) \tag{6.1}$$

where $f = (f_1, f_2, \ldots, f_n)$, and control variables $u = (u_1, u_2, \ldots, u_r)$ and $v = (v_1, v_2, \ldots, v_s)$ are vectors of Euclidean spaces E^r and E^s, respectively. Again we shall assume that x, u, and v belong to fixed domains G, U, and V of E^n, E^r, and E^s, respectively, and that the function f is of class C^1 on $G \times U \times V$.

We shall consider strategies for players J_P and J_E to be functions of x, $p\colon\ x \to p(x)$ and $e\colon\ x \to e(x)$, $x \in G$, respectively, belonging to prescribed classes of functions, and such that:

$$p(x) \in K_u(x) \subseteq U$$
$$e(x) \in K_v(x) \subseteq V$$

where $K_u(x)$ and $K_v(x)$ are given subsets of E^r and E^s, respectively. As indicated by the notation, these subsets may depend on x.

As in Chapter 3, we shall denote by $\mathcal{S}_P$ and $\mathcal{S}_E$ the sets of strategies for J_P and J_E, respectively, and we shall require that $\mathcal{S}_P$ and $\mathcal{S}_E$ satisfy Assumption 3.1.

Mapping γ will be defined as in Example 1.1 of Section 1.2.

Targets θ_P and θ_E will be prescribed *closed*† sets in G. We shall suppose that $\theta_P \cap \theta_E = \varnothing$.

† In the topology induced by E^n on G (see Appendix).

In two target games we shall suppose that $\theta_P \neq \varnothing$ and $\theta_E \neq \varnothing$, and in one target games that $\theta_P \neq \varnothing$ and $\theta_E = \varnothing$. For these two cases, the prescriptions for players J_P and J_E were specified in Chapter 1. Some of the following arguments will apply to both cases.

6.2 A BASIC PROPERTY

Let us note first that if $x \in \partial X$, where X is a subset of G, then for every neighborhood $\Delta(x)$ in G of x

$$\Delta(x) \cap X \neq \varnothing$$

For, if we suppose that $\Delta(x) \cap X = \varnothing$, then

$$x \notin \bar{X}$$

which implies that

$$x \notin \bar{X} \cap \overline{\text{comp}\, X} \triangleq \partial X$$

But this is contrary to our assumption.

Unless otherwise indicated we shall let x^0 denote a point of G which does not belong to $\theta_P \cup \theta_E$, but belongs to the boundary ∂S_P of S_P (or ∂S_E of S_E) and possesses

PROPERTY 6.1. For each given point $x^0 \in \partial S_P$ (or $x^0 \in \partial S_E$) there exist a neighborhood $\Delta(x^0)$ in G of x^0 and a strategy $p^*_{x^0}$ (or $e^*_{x^0}$), such that $p^*_{x^0}$ (or $e^*_{x^0}$) is optimal on $\Delta(x^0) \cap S_P$ (or $\Delta(x^0) \cap S_E$).

We shall let $\{x^0\}$ denote the set of all points x^0 which possess Property 6.1.

6.3 A VARIATIONAL EQUATION AND ITS ADJOINT

Consider a point $x^0 \in \Xi$.† If there exists a neighborhood of x^0 on which $p^*_{x^0}$ and $e^*_{x^0}$ are of class C^1, then there exists a non-null path π_{pe}, emanating from x^0, generated by strategy pair $(p^*_{x^0}, e^*_{x^0})$. Let it be represented by $\mathbf{x}$: $\tau \to \mathbf{x}(\tau)$, $\tau \in [0, \tau_s]$.

Function $\mathbf{x}$ is a solution of

$$dx/d\tau = f(x, p^*_{x^0}(x), e^*_{x^0}(x)) \tag{6.2}$$

with initial condition $\mathbf{x}(0) = x^0$.

† Recall that $\Xi \triangleq \partial S_P \cap \partial S_E$.

If there exists a neighborhood of π_{pe} on which $p^*_{x^0}$ and $e^*_{x^0}$ are of class C^1, we shall associate with Eq. (6.2) the *variational equation*

$$\frac{d\eta}{d\tau} = \left(\frac{\partial f}{\partial x} + \frac{\partial f}{\partial u}\frac{\partial p^{*_0}_x}{\partial x} + \frac{\partial f}{\partial v}\frac{\partial e^{*_0}_x}{\partial x}\right)\Bigg|_{x=\mathbf{x}(\tau)} \eta \tag{6.3}$$

where

$$\frac{\partial f}{\partial x} \triangleq \left[\frac{\partial f_\nu(x, u, v)}{\partial x_\alpha}\right], \qquad \begin{aligned}\nu &= 1, 2, \ldots, n\\ \alpha &= 1, 2, \ldots, n\end{aligned}$$

$$\frac{\partial f}{\partial u} \triangleq \left[\frac{\partial f_\nu(x, u, v)}{\partial u_\alpha}\right], \qquad \begin{aligned}\nu &= 1, 2, \ldots, n\\ \alpha &= 1, 2, \ldots, r\end{aligned}$$

$$\frac{\partial f}{\partial v} \triangleq \left[\frac{\partial f_\nu(x, u, v)}{\partial v_\alpha}\right], \qquad \begin{aligned}\nu &= 1, 2, \ldots, n\\ \alpha &= 1, 2, \ldots, s\end{aligned}$$

and

$$\frac{\partial p}{\partial x} \triangleq \left[\frac{\partial p_\nu(x)}{\partial x_\alpha}\right], \qquad \begin{aligned}\nu &= 1, 2, \ldots, r\\ \alpha &= 1, 2, \ldots, n\end{aligned}$$

with $p = p^*_{x^0}$,

$$\frac{\partial e}{\partial x} \triangleq \left[\frac{\partial e_\nu(x)}{\partial x_\alpha}\right], \qquad \begin{aligned}\nu &= 1, 2, \ldots, s\\ \alpha &= 1, 2, \ldots, n\end{aligned}$$

with $e = e^*_{x^0}$; and

$$\eta \triangleq \begin{pmatrix}\eta_1\\ \eta_2\\ \cdot\\ \cdot\\ \cdot\\ \eta_n\end{pmatrix}$$

By definition, the equation *adjoint* to Eq. (6.3) is

$$\frac{d\lambda}{d\tau} = -\left(\frac{\partial f}{\partial x} + \frac{\partial f}{\partial u}\frac{\partial p^{*_0}_x}{\partial x} + \frac{\partial f}{\partial v}\frac{\partial e^{*_0}_x}{\partial x}\right)^{\mathrm{T}}\Bigg|_{x=\mathbf{x}(\tau)} \lambda \tag{6.4}$$

where

$$\lambda \triangleq \begin{pmatrix}\lambda_1\\ \lambda_2\\ \cdot\\ \cdot\\ \cdot\\ \lambda_n\end{pmatrix}$$

For given initial conditions $\eta = \eta^0$ and $\lambda = \lambda^0$ at $\tau = 0$, the solutions $\boldsymbol{\eta}$: $\tau \to \eta = \boldsymbol{\eta}(\tau)$ and $\boldsymbol{\lambda}$: $\tau \to \lambda = \boldsymbol{\lambda}(\tau)$ of Eq. (6.3) and Eq. (6.4), respectively, are unique and continuous on $[0, \tau_s]$. Furthermore, $\boldsymbol{\eta}(\tau)$ and $\boldsymbol{\lambda}(\tau)$ are nonzero provided that η^0 and λ^0, respectively, are nonzero.

6.4 A PROPERTY OF BOUNDARIES OF SETS OF THE GAME

We shall now prove

Lemma 6.1. *Let* $\mathbf{x}$*:* $\tau \to \mathbf{x}(\tau)$, $\tau \in [0, \tau_s]$, *represent a path* π_p, *emanating from point* $x^0 = \mathbf{x}(0)$, $x^0 \in \partial S_P$, *generated by strategy pair* $(p^*_{x^0}, e)$, $e \in \mathscr{S}_E$. *If at every point of* π_p, *with the possible exception of its end point, there exists a neighborhood on which* $p^*_{x^0}$ *and* e *are of class* C^1, *then* $\pi_p \subset \bar{S}_P$.

If π_p is a null path, Lemma 6.1 is trivial. Hence, let us suppose that π_p is not a null path.

First of all, let us note that, since θ_P and θ_E are closed, and since $\mathbf{x}(\tau) \neq \theta_P \cup \theta_E$ for all $\tau \in [0, \tau_s)$, there exists at every point $\mathbf{x}(\tau)$,

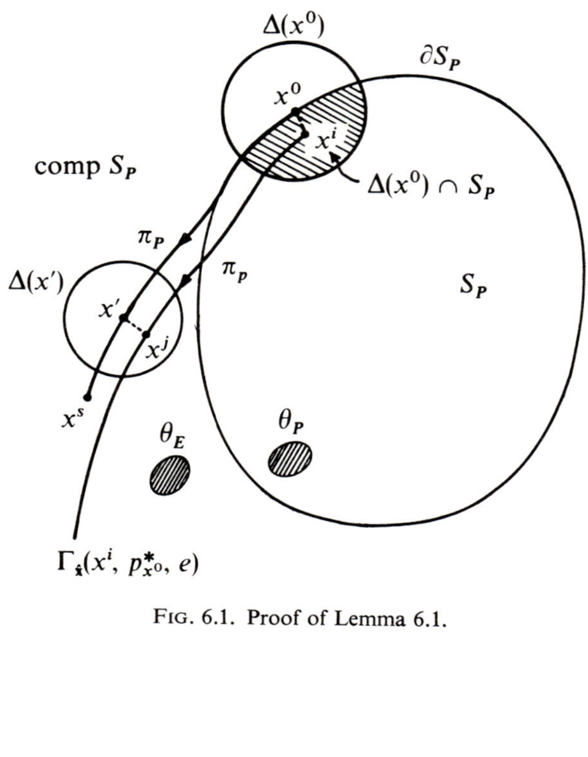

FIG. 6.1. Proof of Lemma 6.1.

$\tau \in [0, \tau_s)$, a neighborhood $\Delta(\mathbf{x}(\tau))$ in G of $\mathbf{x}(\tau)$ such that

$$\Delta(\mathbf{x}(\tau)) \cap (\theta_P \cup \theta_E) = \varnothing$$

Let $x' = \mathbf{x}(\tau')$, $\tau' \in [0, \tau_s)$, and let us suppose that $\mathbf{x}(\tau') \notin S_P$; that is,

$$\mathbf{x}(\tau') \in \overset{\circ}{\widehat{\text{comp } S_P}}$$

Let $\Delta(\mathbf{x}(\tau'))$ be a neighborhood in G of $\mathbf{x}(\tau')$ such that

$$\Delta(\mathbf{x}(\tau')) \subset \overset{\circ}{\widehat{\text{comp } S_P}}$$

Let $\Delta(x^0)$ be a neighborhood of x^0 such that $p^*_{x^0}$ is optimal on $\Delta(x^0) \cap S_P$, and consider a point of $\Delta(x^0) \cap S_P$, namely,†

$$x^i \triangleq x^0 + \epsilon\eta^0 \in \Delta(x^0) \cap S_P$$

Now let us consider describing curve $\Gamma_{\hat{\mathbf{x}}}(x^i, p^*_{x^0}, e)$ that emanates from x^i, is generated by strategy pair $(p^*_{x^0}, e)$, and is given by $\hat{\mathbf{x}} \in \gamma(x^i, p^*_{x^0}, e)$.

Since at every point of π_p, with the possible exception of its terminal point, there exists a neighborhood on which $p^*_{x^0}$ and e are of class C^1, then, for $|\epsilon|$ sufficiently small, $\hat{\mathbf{x}}$ is a solution of trajectory equation

$$dx/d\tau = f(x, p^*_{x^0}(x), e(x))$$

With initial condition $\hat{\mathbf{x}}(0) = x^i$.

Furthermore, for $|\epsilon|$ sufficiently small, there exists a point of $\Gamma_{\hat{\mathbf{x}}}(x^i, p^*_{x^0}, e)$ corresponding to $\tau = \tau'$, namely

$$x^j \triangleq \hat{\mathbf{x}}(\tau') = x' + \epsilon\eta' + o(\tau', \epsilon)$$

where $\lim_{\epsilon \to 0} \|o(\tau', \epsilon)\|/\epsilon = 0$, $\eta' = \boldsymbol{\eta}(\tau')$, and $\boldsymbol{\eta}\colon\ \tau \to \eta = \boldsymbol{\eta}(\tau)$ is the solution of linear equation

$$\frac{d\eta}{d\tau} = \left(\frac{\partial f}{\partial x} + \frac{\partial f}{\partial u}\frac{\partial p^*_{x^0}}{\partial x} + \frac{\partial f}{\partial v}\frac{\partial e}{\partial x}\right)\bigg|_{x=\mathbf{x}(\tau)} \eta$$

defined on $[0, \tau_s]$, subject to the condition $\boldsymbol{\eta}(0) = \eta^0$.

Since at every point $\mathbf{x}(\tau)$, $\tau \in [0, \tau_s)$, there exists a neighborhood $\Delta(\mathbf{x}(\tau))$ in G such that

$$\Delta(\mathbf{x}(\tau)) \cap (\theta_P \cup \theta_E) = \varnothing$$

† If there exists no point other than x^0 in $\Delta(x^0) \cap S_P$, then $x^0 \in S_P$. In that case, Lemma 6.1 reduces to Lemma 5.1.

then, for $|\epsilon|$ sufficiently small, $\hat{\mathbf{x}}(\tau) \notin \theta_P \cup \theta_E$ for all $\tau \in [0, \tau']$; and hence, trajectory Γ^{ij} given by $\hat{\mathbf{x}}$: $\tau \to \hat{\mathbf{x}}(\tau)$, $\tau \in [0, \tau']$, is a path, say $\hat{\pi}_p$.

Furthermore, for $|\epsilon|$ sufficiently small,

$$x^j \triangleq \hat{\mathbf{x}}(\tau') \in \Delta(\mathbf{x}(\tau')) \subset \widehat{\overset{\circ}{\text{comp}}\, S_P} \subset \text{comp}\, S_P \tag{6.5}$$

Since $p^*_{x^0}$ is optimal on $\Delta(x^0) \cap S_P$, it is optimal at point x^i, and hence $\hat{\pi}_p$ is a $\mathscr{P}$-path; namely,

$$\hat{\pi}_p = \pi^{ij}_{\mathscr{P}} \tag{6.6}$$

But (6.5) and (6.6) contradict Lemma 5.1, and hence

$$\mathbf{x}(\tau') \in \bar{S}_P \tag{6.7}$$

Since (6.7) holds for all $\tau' \in [0, \tau_s)$, and since $\mathbf{x}$ is continuous on $[0, \tau_s]$,

$$\mathbf{x}(\tau_s) \in \bar{S}_P$$

And so Lemma 6.1 is established.

By similar arguments one can also prove

Lemma 6.2. *Let* $\mathbf{x}$: $\tau \to \mathbf{x}(\tau)$, $\tau \in [0, \tau_s]$, *represent a path* π_e, *emanating from point* $x^0 = \mathbf{x}(0)$, $x^0 \in \partial S_E$, *generated by strategy pair* $(p, e^*_{x^0})$, $p \in \mathscr{S}_P$. *If at every point of* π_e, *with the possibility exception of its end point, there exists a neighborhood on which* p *and* $e^*_{x^0}$ *are of class* C^1, *then* $\pi_e \subset \bar{S}_E$.

6.5 A PROPERTY OF THE SURFACE OF THE GAME

As a direct consequence of Lemmas 6.1 and 6.2 we have

Corollary 6.1. *Let* $\mathbf{x}$: $\tau \to \mathbf{x}(\tau)$, $\tau \in [0, \tau_s]$, *represent a path* π_{pe}, *emanating from point* $x^0 = \mathbf{x}(0)$, $x^0 \in \Xi$, *generated by strategy pair* $(p^*_{x^0}, e^*_{x^0})$. *If at every point of* π_{pe}, *with the possible exception of its end point, there exists a neighborhood on which* $p^*_{x^0}$ *and* $e^*_{x^0}$ *are of class* C^1 *then* $\pi_{pe} \subset \Xi$.

From the definition of Ξ it follows that x^0 belongs to both ∂S_P and ∂S_E.

As a consequence of Lemma 6.1,

$$\pi_{pe} \subset \bar{S}_P$$

and, by Lemma 6.2,

$$\pi_{pe} \subset \bar{S}_E$$

Hence,

$$\pi_{pe} \subset \bar{S}_P \cap \bar{S}_E \tag{6.8}$$

Finally, (6.8) and (5.11) imply that

$$\pi_{pe} \subset \Xi$$

We see that Lemmas 6.1, 6.2, and Corollary 6.1 exhibit strong similarities to Lemmas 2.1, 2.2 and Corollary 2.1, respectively. They enforce the remarks of Section 5.5. It appears now that, provided the assumptions of these lemmas and corollaries are satisfied, surface Ξ has limiting properties similar to the ones of a surface $\Sigma(C)$; that is, surface Ξ is the common boundary of two regions which contain all paths π_p and π_e, respectively, emanating from $\{x^0\} \subseteq \Xi$.

Definition 6.1. A path π_{pe} emanating from a point $x^0 \in \Xi$, generated by a strategy pair $(p^*_{x^0}, e^*_{x^0})$, will be called a *boundary path.*

6.6 TWO FAMILIES OF PATHS

Let us consider a point $x^0 \in \Xi$, and assume that there exist strategies $p^*_{x^0}$ and $e^*_{x^0}$ which are of class C^1 on some neighborhood of x^0. Then there exists a non-null path π_{pe}, emanating from x^0, generated by $(p^*_{x^0}, e^*_{x^0})$, and given by $\mathbf{x}$: $\tau \to x = \mathbf{x}(\tau)$, $\tau \in [0, \tau_s]$.

Next we introduce

Assumption 6.1. For all $x \in \pi_{pe}$, and for all $u \in K_u(x)$ and $v \in K_v(x)$, there exist strategies $\alpha_x \in \mathscr{S}_P$ and $\beta_x \in \mathscr{S}_E$ which are of class C^1 on some neighborhood of x, and which satisfy the conditions $\alpha_x(x) = u$ and $\beta_x(x) = v$, respectively.

Assumption 6.2. At every point x of π_{pe}, there exists a neighborhood $\Delta(x)$ on which $p^*_{x^0}$ and $e^*_{x^0}$ are of class C^1.

As a consequence of Corollary 6.1 it follows that $\pi_{pe} \subset \Xi$.

Let $x' = \mathbf{x}(\tau')$, $\tau' \in [0, \tau_s]$, be a point of π_{pe} that does not belong to $\theta_P \cup \theta_E$. Of course, $x' \in \Xi$. Now consider paths π_p' and π_e' emanating from $x' = \mathbf{x}(\tau')$, generated by strategy pairs $(p^*_{x^0}, \beta_{x'})$ and $(\alpha_{x'}, e^*_{x^0})$,

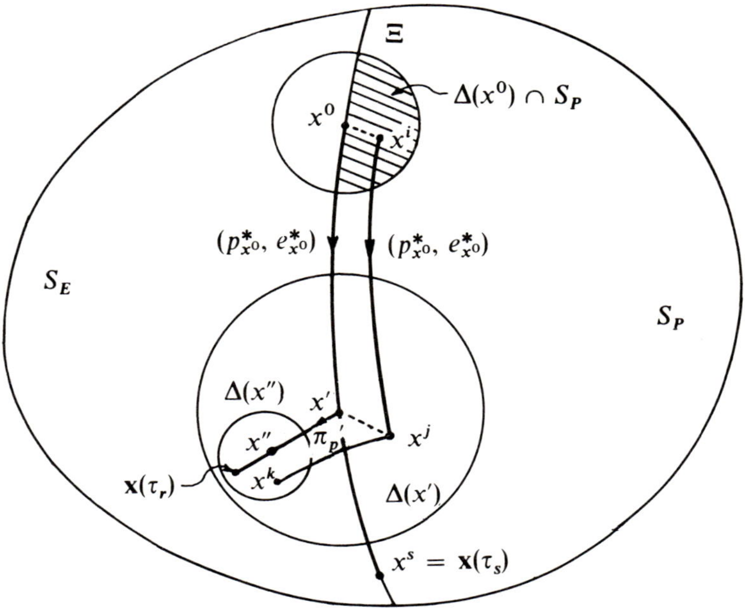

FIG. 6.2. Proof of the basic property of paths π_p'.

respectively, where

$$\alpha_{x'}(x') = u^b \in K_u(x')$$
$$\beta_{x'}(x') = v^b \in K_v(x')$$

and u^b, v^b are given vectors.

Let us suppose that π_p' and π_e' belong to a neighborhood $\Delta(x')$ on which $p^*_{x^0}$, $\beta_{x'}$ and $\alpha_{x'}$, $e^*_{x^0}$ are of class C^1. We shall prove that $\pi_p' \subset \bar{S}_P$ and $\pi_e' \subset \bar{S}_E$.

These properties are trivial in the case of null paths; consequently let us suppose that π_p' and π_e' are non-null paths. Such paths exist since $p^*_{x^0}$, $\beta_{x'}$ and $\alpha_{x'}$, $e^*_{x^0}$ are of class C^1 on $\Delta(x')$, and since $x' \notin \theta_P \cup \theta_E$.

For instance, consider path π_p' and let it be represented by $\mathbf{y}$: $\tau \to \mathbf{y}(\tau)$, $\tau \in [0, \tau_r]$.

Let $x'' = \mathbf{y}(\tau'')$, $\tau'' \in [0, \tau_r)$, and let us suppose that $x'' \notin \bar{S}_P$; that is,

$$x'' \in \overset{\circ}{\widehat{\operatorname{comp} S_P}}$$

Let $\Delta(x'')$ be a neighborhood in G of $\mathbf{y}(\tau'')$ such that

$$\Delta(x'') \subset \overset{\circ}{\widehat{\text{comp } S_P}} \cap \Delta(x')$$

Let $\Delta(x^0)$ be a neighborhood of x^0 such that $p^*_{x^0}$ is optimal on $\Delta(x^0) \cap S_P$, and let

$$x^i \triangleq x^0 + \epsilon\eta^0 \in \Delta(x^0) \cap S_P$$

By arguments similar to those employed in the proof of Lemma 6.1, one can readily verify that, for $|\epsilon|$ sufficiently small, there exists a path that emanates from x^i, is generated by $(p^*_{x^0}, e^*_{x^0})$, and ends at point x^j at $\tau = \tau'$. Since $p^*_{x^0}$ is optimal at point x^i, that path is a $\mathscr{P}$-path, $\pi^{ij}_{\mathscr{P}}$. Let it be represented by $\hat{\mathbf{x}}$: $\tau \to \hat{\mathbf{x}}(\tau)$, $\tau \in [0, \tau']$.

Point $x^j \triangleq \hat{\mathbf{x}}(\tau')$ is given by

$$x^j = x' + \epsilon\eta' + o(\tau', \epsilon)$$

where $\lim_{\epsilon \to 0} \|o(\tau', \epsilon)\|/\epsilon = 0$, $\eta' = \boldsymbol{\eta}(\tau')$, and $\boldsymbol{\eta}$: $\tau \to \eta = \boldsymbol{\eta}(\tau)$, $\tau \in [0, \tau_s]$ is the solution of variational equation (6.3) subject to the condition $\boldsymbol{\eta}(0) = \eta^0$.

It follows from Lemma 5.1 that

$$x^j \in S_P$$

and, for $|\epsilon|$ sufficiently small,

$$x^j \in S_P \cap \Delta(x')$$

For $|\epsilon|$ sufficiently small, there also exists a path π^{jk} that emanates from x^j, is generated by $(p^*_{x^0}, \beta_{x'})$, ends at point x^k at $\tau = \tau''$, and belongs to $\Delta(x')$. Let it be represented by $\hat{\mathbf{y}}$: $\tau \to \hat{\mathbf{y}}(\tau)$, $\tau \in [0, \tau'']$.

Since x^j is arbitrarily close to x^0, x^j is arbitrarily close to x', and since $p^*_{x^0}$ and $\beta_{x'}$ are of class C^1 on $\Delta(x')$, $x^k \triangleq \hat{\mathbf{y}}(\tau'')$ is arbitrarily close to x''; that is, for $|\epsilon|$ sufficiently small,

$$x^k \in \Delta(x'')$$

and hence

$$x^k \in \overset{\circ}{\widehat{\text{comp } S_P}}, \tag{6.9}$$

In this manner we have generated

$$\pi^{ij}_{\mathscr{P}} \cup \pi^{jk}$$

which can be represented by the function $\boldsymbol{\zeta}$: $\tau \to \boldsymbol{\zeta}(\tau)$, $\tau \in [0, \tau' + \tau'']$, where

$$\boldsymbol{\zeta}(\tau) = \hat{\mathbf{x}}(\tau) \qquad \text{for} \qquad \tau \in [0, \tau')$$

$$\boldsymbol{\zeta}(\tau) = \hat{\mathbf{y}}(\tau - \tau') \qquad \text{for} \qquad \tau \in [\tau', \tau' + \tau'']$$

Employing arguments similar to those used in Section 3.8, one can readily verify that $\pi_{\mathscr{P}}^{ij} \cup \pi^{jk}$ is a path that emanates from x^i and is generated by strategy pair $(p_{x^0}^*, \hat{e})$, where

$$\hat{e}(x) = e_{x^0}^*(x) \qquad \text{for all } x \in G \text{ such that } x_n < x_n'$$

$$\hat{e}(x) = \beta_{x'}(x) \qquad \text{for all } x \in G \text{ such that } x_n \geqslant x_n'$$

That is, $\pi_{\mathscr{P}}^{ij} \cup \pi^{jk}$ is a $\mathscr{P}$-path, say $\pi_{\mathscr{P}}^{ik}$.

However, (6.9) contradicts Lemma 5.1, and hence

$$x'' \in \bar{S}_P \tag{6.10}$$

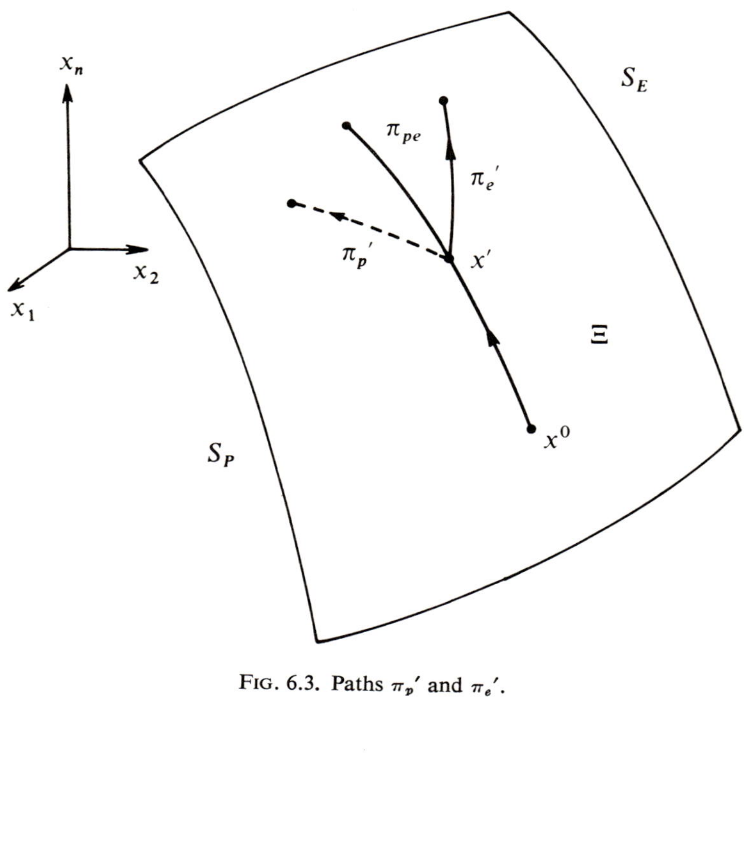

FIG. 6.3. Paths π_p' and π_e'.

Since (6.10) holds for all $\tau'' \in [0, \tau_r)$, and since $\mathbf{y}$ is continuous on $[0, \tau_r]$,

$$\mathbf{y}(\tau_r) \in \bar{S}_P$$

Hence we have proved that

$$\pi_p' \subset \bar{S}_P \tag{6.11}$$

By similar arguments one can establish that

$$\pi_e' \subset \bar{S}_E \tag{6.12}$$

Finally, if $\tau' = \tau_s$ and $x' = \mathbf{x}(\tau_s) \in \theta_P \cup \theta_E$, then π_p' and π_e' are null paths for which (6.11) and (6.12) are obviously valid.

Next we shall consider two families of paths, namely

1. the set $\{\pi_p'\}$ of all paths π_p' emanating from points $x' \in \pi_{pe}$; and
2. the set $\{\pi_e'\}$ of all paths π_e' emanating from points $x' \in \pi_{pe}$.

6.7 A LOCAL MIN-MAX CONDITION

Hereafter we shall utilize the notation and the assumptions of Section 6.6. In addition, we shall introduce

Assumption 6.3. Game surface Ξ can be defined by the equation

$$\mathbf{g}(x) = 0$$

and at each point $x' = \mathbf{x}(\tau')$, $x' \notin \theta_P \cup \theta_E$, of a path π_{pe} starting at $x^0 \in \Xi$, there exists a neighborhood $\Delta(x')$ in G of x' on which the function $\mathbf{g}$ is of class C^1 and grad $\mathbf{g}(x)$ is defined and different from zero. Furthermore at every point x of $\Delta(x') \cap \bar{S}_P$,

$$\mathbf{g}(x) \leqslant 0 \tag{6.13}$$

and at every point x of $\Delta(x') \cap \bar{S}_E$,

$$\mathbf{g}(x) \geqslant 0 \tag{6.14}$$

Again, consider a non-null path π_e' emanating from x', $x' \notin \theta_P \cup \theta_E$, generated by strategy pair $(\alpha_{x'}, e^*_{x^0})$.†

Let $x' + \delta^e x$ denote points of π_e' in the neighborhood $\Delta(x')$, and let ξ be a function

$$\xi\colon \quad \tau \to \delta^e x = \xi(\tau)$$

where $\xi(0) = 0$ at point x', defined on an interval that depends on $\Delta(x')$.

† We assume that $\alpha_{x'}$ is of class C^1 on $\Delta(x')$ and that $\pi_e' \subset \Delta(x')$.

It follows from (6.12) and (6.14) that

$$\mathbf{g}(x' + \delta^e x) \geqslant 0$$

and $\|\delta^e x\| \to 0$ as $\tau \to 0$; hence, since $x' \in \Xi$ and accordingly $\mathbf{g}(x') = 0$,

$$\operatorname{grad} \mathbf{g}(x') \cdot \delta^e x + o(\|\delta^e x\|) \geqslant 0 \tag{6.15}$$

where

$$\lim_{\|\delta^e x\| \to 0} o(\|\delta^e x\|)/\|\delta^e x\| = 0$$

Upon dividing (6.15) by τ, which is positive for $\delta^e x \neq 0$, we obtain

$$\operatorname{grad} \mathbf{g}(x') \cdot \frac{\delta^e x}{\tau} + \frac{o(\|\delta^e x\|)}{\tau} \geqslant 0$$

Then allowing $\tau \to 0$, and consequently

$$\delta^e x/\tau \to f(x', u^b, e^*_{x0}(x')) \qquad \text{where} \qquad u^b = \alpha_{x'}(x')$$

leads to

$$\operatorname{grad} \mathbf{g}(x') \cdot f(x', u^b, e^*_{x0}(x')) \geqslant 0$$
$$\forall x', \quad x' \in \pi_{pe}, \quad x' \notin \theta_P \cup \theta_E \qquad \forall u^b \in K_u(x') \tag{6.16}$$

Invoking similar arguments for a path π_p' emanating from $x' \in \pi_{pe}$, $x' \notin \theta_P \cup \theta_E$, generated by strategy pair $(p^*_{x0}, \beta_{x'})$, results in

$$\operatorname{grad} \mathbf{g}(x') \cdot f(x', p^*_{x0}(x'), v^b) \leqslant 0$$
$$\forall x', \quad x' \in \pi_{pe}, \quad x' \notin \theta_P \cup \theta_E \qquad \forall v^b \in K_v(x') \tag{6.17}$$

Moreover, since $\pi_{pe} \subset \Xi$,

$$\operatorname{grad} \mathbf{g}(x') \cdot f(x', p^*_{x0}(x'), e^*_x(x')) = 0$$
$$\forall x', \quad x' \in \pi_{pe}, \quad x' \notin \theta_P \cup \theta_E \tag{6.18}$$

Conditions (6.16)–(6.18) are embodied in

Lemma 6.3. *Let π_{pe} be a boundary path. If Assumptions* 6.1–6.3 *are satisfied, then at every point x of π_{pe} which does not belong to $\theta_P \cup \theta_E$,*

$$\min_{u \in K_u(x)} \mathbf{n}(x) \cdot f(x, u, e^*_{x0}(x))$$
$$= \max_{v \in K_v(x)} \mathbf{n}(x) \cdot f(x, p^*_{x0}(x), v)$$
$$= \mathbf{n}(x) \cdot f(x, p^*_{x0}(x), e^*_{x0}(x)) = 0 \tag{6.19}$$

where $\mathbf{n}(x) \triangleq \text{grad}\, \mathbf{g}(x)$, x^0 *is the initial point of* π_{pe}, *and* $(p^*_{x^0}, e^*_{x^0})$ *is the strategy pair that generates* π_{pe}.

6.8 GRADIENT AND ADJOINT VECTOR

Consider a path π_{pe} that emanates from $x^0 \in \Xi$, is generated by $(p^*_{x^0}, e^*_{x^0})$ and is given by function $\mathbf{x}$: $\tau \to x = \mathbf{x}(\tau)$, $\tau \in [0, \tau_s]$.

Let $\Delta(x^0)$ be a neighborhood in G of x^0 such that $p^*_{x^0}$ and $e^*_{x^0}$ are optimal on $\Delta(x^0) \cap S_P$ and $\Delta(x^0) \cap S_E$, respectively, and such that

$$\Delta(x^0) \cap (\theta_P \cup \theta_E) = \varnothing.$$

Let $\hat{x}^0$ denote a point of $\Delta(x^0) \cap \Xi$.† Since grad $\mathbf{g}(x^0)$ is defined, Ξ is such that in arbitrarily small ball $\Delta'(x^0)$, with center x^0, there exist $n - 1$ points $\hat{x}^0$ which may be represented by

$$\hat{x}^0 = x^0 + \epsilon\eta^0 + o(\epsilon) \tag{6.20}$$

where $\lim_{\epsilon \to 0} \|o(\epsilon)\|/\epsilon = 0$ and grad $\mathbf{g}(x^0) \cdot \eta^0 = 0$, and for which $|\epsilon| \neq 0$ and the $n - 1$ vectors η^0 are linearly independent.

Now, with each point $\hat{x}^0$ let us associate a path $\hat{\pi}_{pe}$ that emanates from it and is generated by $(p^*_{x^0}, e^*_{x^0})$.

The points of such paths, say $\hat{x}'$, at $\tau = \tau'$, $\tau' > 0$, *provided that they exist*, are obtained by integrating Eq. (6.2) subject to the initial condition $x = \hat{x}^0$ at $\tau = 0$.

Provided that $x' = \mathbf{x}(\tau')$ does not belong to $\theta_P \cup \theta_E$, that Assumption 6.2 is satisfied along π_{pe}, and that $|\epsilon|$ is sufficiently small, points $\hat{x}'$ exist and are given by

$$\hat{x}' = x' + \epsilon\eta' + o(\epsilon)$$

with $\eta' = \boldsymbol{\eta}(\tau')$, where $\boldsymbol{\eta}$ is the solution of variational equation (6.3) subject to the initial condition $\boldsymbol{\eta}(0) = \eta^0$.

Corollary 6.1 implies that

$$\hat{x}' = x' + \epsilon\eta' + o(\epsilon) \in \Xi$$

Furthermore,

$$\mathbf{g}(\hat{x}') - \mathbf{g}(x') = \text{grad}\, \mathbf{g}(x') \cdot (\epsilon\eta' + o(\epsilon)) + o(\|\epsilon\eta' + o(\epsilon)\|)$$

or, more simply,

$$\mathbf{g}(\hat{x}') - \mathbf{g}(x') = \epsilon \,\text{grad}\, \mathbf{g}(x') \cdot \eta' + o(\epsilon)$$

where $\lim_{\epsilon \to 0} o(\epsilon)/\epsilon = 0$.

† Clearly, $\hat{x}^0$ belongs to the set $\{x^0\}$ of all points for which Property 6.1 is satisfied.

Moreover, since x' and $\hat{x}'$ belong to Ξ, $\mathbf{g}(x') = \mathbf{g}(\hat{x}') = 0$, and hence

$$\epsilon \text{ grad } \mathbf{g}(x') \cdot \eta' + o(\epsilon) = 0 \tag{6.21}$$

Upon dividing (6.21) by ϵ, and letting $\epsilon \to 0$, we arrive at

$$\text{grad } \mathbf{g}(x') \cdot \eta' = 0 \tag{6.22}$$

This is true for $n - 1$ linearly independent vectors η' and for *all* transfer times $\tau' \in [0, \tau_s)$.

Condition (6.22) can be written as

$$\text{grad } \mathbf{g}(\mathbf{x}(\tau)) \cdot \boldsymbol{\eta}(\tau) = 0 \qquad \tau \in [0, \tau_s) \tag{6.23}$$

Now consider the solution $\boldsymbol{\lambda}$: $\tau \to \lambda = \boldsymbol{\lambda}(\tau)$, $\tau \in [0, \tau_s]$, of adjoint equation (6.4) for given initial condition $\boldsymbol{\lambda}(0) = \lambda^0$. This vector has the property that

$$(d/d\tau)(\boldsymbol{\lambda}(\tau) \cdot \boldsymbol{\eta}(\tau)) = 0$$

and so, if we choose λ^0 perpendicular to the tangent plane of Ξ at point x^0, that is if λ^0 has the same supporting line as grad $\mathbf{g}(x^0)$, then $\boldsymbol{\lambda}(\tau)$ is perpendicular to the tangent plane of Ξ at point $\mathbf{x}(\tau)$ for all τ for which $\mathbf{x}(\tau) \in \pi_{pe}$, with the possible exception of the terminal point of π_{pe}; that is, its supporting line is that of grad $\mathbf{g}(\mathbf{x}(\tau))$.

Function $\boldsymbol{\lambda}$ is a nonzero continuous vector function, and so, if we choose $\boldsymbol{\lambda}(\tau)$ in the same direction as grad $\mathbf{g}(\mathbf{x}(\tau))$ at $\tau = 0$, it remains codirectional with grad $\mathbf{g}(\mathbf{x}(\tau))$ for all τ for which $\mathbf{x}(\tau) \in \pi_{pe}$, with the possible exception of the terminal point of π_{pe}.

6.9 TRANSVERSALITY CONDITION

Again consider a path π_{pe} that emanates from $x^0 \in \Xi$, is generated by $(p^*_{x^0}, e^*_{x^0})$ and is given by function $\mathbf{x}$.

Let $\Delta(x^0)$ be a neighborhood in G of x^0 such that $p^*_{x^0}$ and $e^*_{x^0}$ are optimal on $\Delta(x^0) \cap S_P$ and $\Delta(x^0) \cap S_E$, respectively, and such that, at every point of $\Delta(x^0) \cap \bar{S}_P$,

$$\mathbf{g}(x) \leqslant 0 \tag{6.24}$$

and at every point of $\Delta(x^0) \cap \bar{S}_E$,

$$\mathbf{g}(x) \geqslant 0 \tag{6.25}$$

In addition to Assumption 6.3, we shall make

Assumption 6.4. $\Delta(x^0) \cap \overset{\circ}{S}_P$ and $\Delta(x^0) \cap \bar{S}_E$ (or $\Delta(x^0) \cap \bar{S}_P$ and $\Delta(x^0) \cap \overset{\circ}{S}_E$) constitute a *partition* of $\Delta(x^0)$.

Here we shall study the case for which the end point of π_{pe}, namely $x^f \triangleq \mathbf{x}(\tau_f)$, belongs either to θ_P or to θ_E.

Next we introduce

Assumption 6.5. The boundary of the target to which x^f belongs, namely

$$\partial\theta_P \triangleq \bar{\theta}_P \cap \overline{\text{comp}\,\theta_P}$$

or

$$\partial\theta_E \triangleq \bar{\theta}_E \cap \overline{\text{comp}\,\theta_E}$$

is defined by the equation

$$\boldsymbol{\theta}(x) = 0 \tag{6.26}$$

where $\boldsymbol{\theta} = \boldsymbol{\theta}_P$ or $\boldsymbol{\theta} = \boldsymbol{\theta}_E$, respectively.

Furthermore, there exists a neighborhood $\Delta(x^f)$ of x^f on which the function $\boldsymbol{\theta}$ is of class C^1 and grad $\boldsymbol{\theta}(x)$ is defined and different from zero.

Let $\hat{x}^f$ denote a point of $\Delta(x^f) \cap \partial\theta_P$ (or of $\Delta(x^f) \cap \partial\theta_E$). Since grad $\boldsymbol{\theta}(x^f)$ is defined, $\partial\theta_P$ (or $\partial\theta_E$) is such that in an arbitrarily small ball $\Delta'(x^f)$, with center x^f, there exist $n - 1$ points $\hat{x}^f$ which may be represented by

$$\hat{x}^f = x^f + \epsilon\eta^f + o(\epsilon) \tag{6.27}$$

where $\lim_{\epsilon\to 0} \|o(\epsilon)\|/\epsilon = 0$, and grad $\boldsymbol{\theta}(x^f) \cdot \eta^f = 0$, and for which ϵ is strictly positive and the $n - 1$ vectors η^f are linearly independent. Moreover, to each point $\hat{x}^f$ for which ϵ is positive there corresponds a point in $\Delta'(x^f) \cap \partial\theta_P$ (or $\Delta'(x^f) \cap \partial\theta_E$) for which ϵ is negative and η^f is the same.

For instance, suppose that $x^f \in \theta_P$. Let λ^0 be perpendicular to the tangent plane of Ξ at point x^0 and in the same direction as grad $\mathbf{g}(\mathbf{x}(0))$. Let $\lambda^f \triangleq \boldsymbol{\lambda}(\tau_f)$, where $\boldsymbol{\lambda}$ is the solution of adjoint equation (6.4) with $\boldsymbol{\lambda}(0) = \lambda^0$.

We shall prove that

$$\lambda^f \cdot \eta^f = 0 \tag{6.28}$$

for all η^f in (6.27), and hence that λ^f is collinear with grad $\boldsymbol{\theta}(x^f)$.

Let us suppose that $\lambda^f \cdot \eta^f \neq 0$ for some η^f; that is, λ^f is not perpendicular to the tangent plane of θ_P at point x^f.

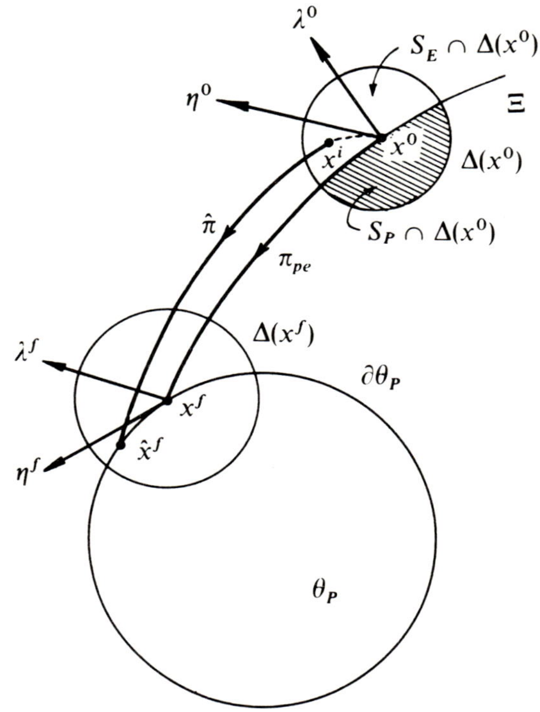

FIG. 6.4. Proof of the transversality condition.

If $\lambda^f \cdot \eta^f > 0$, we let

$$\hat{x}^f = x^f + \epsilon\eta^f + o(\epsilon) \qquad \text{with} \qquad \epsilon > 0$$

If $\lambda^f \cdot \eta^f < 0$, we let

$$\hat{x}^f = x^f + \epsilon\eta^f + o(\epsilon) \qquad \text{with} \qquad \epsilon < 0$$

If Assumption 6.2 is satisfied along π_{pe}, there exists† a solution $\hat{\mathbf{x}}$: $\tau \to x = \hat{\mathbf{x}}(\tau)$, $\tau \in [0, \tau_f]$, of trajectory equation (6.2) that satisfies the end condition

$$\hat{\mathbf{x}}(\tau_f) = \hat{x}^f.$$

and there exists a point $x^i = \hat{\mathbf{x}}(0)$ such that

$$x^i = x^0 + \epsilon\eta^0 + o(\epsilon)$$

† For $|\epsilon|$ sufficiently small.

with $\eta^0 = \boldsymbol{\eta}(0)$, where $\boldsymbol{\eta}$ is the solution of variational equation (6.3) with end condition $\eta^f = \boldsymbol{\eta}(\tau_f)$.

We have

$$\mathbf{g}(x^i) = \mathbf{g}(x^0 + \epsilon\eta^0 + o(\epsilon))$$

and, since $\mathbf{g}(x^0) = 0$,

$$\begin{aligned}\mathbf{g}(x^i) &= \operatorname{grad} \mathbf{g}(x^0) \cdot (\epsilon\eta^0 + o(\epsilon)) + o(\|\epsilon\eta^0 + o(\epsilon)\|) \\ &= \epsilon \operatorname{grad} \mathbf{g}(x^0) \cdot \eta^0 + o(\epsilon)\end{aligned}$$

where $\lim_{\epsilon\to 0} o(\epsilon)/\epsilon = 0$. For $|\epsilon|$ sufficiently small, the sign of the right-hand side is the same as the sign of $\epsilon \operatorname{grad} \mathbf{g}(x^0) \cdot \eta^0$; that is, $\mathbf{g}(x^i)$ has the same sign as $\epsilon\lambda^0 \cdot \eta^0$.

Since $\boldsymbol{\lambda}(\tau) \cdot \boldsymbol{\eta}(\tau) =$ constant along π_{pe}, for $|\epsilon|$ sufficiently small, $\mathbf{g}(x^i)$ has the same sign as $\epsilon\lambda^f \cdot \eta^f$; that is,

$$\mathbf{g}(x^i) > 0 \tag{6.29}$$

Also, for $|\epsilon|$ sufficiently small,

$$x^i \in \Delta(x^0)$$

It follows from (6.24), (6.25), (6.29), and Assumption 6.4 that

$$x^i \in S_E \cap \Delta(x^0)$$

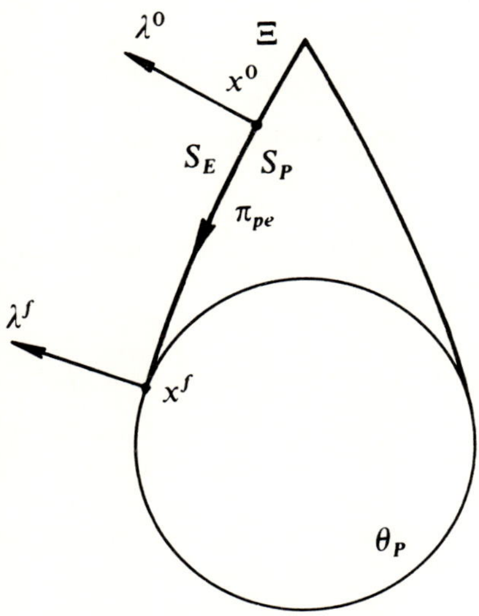

FIG. 6.5. Transversality condition.

Since $\theta_P \cap \theta_E = \varnothing$, for $|\epsilon|$ sufficiently small there does not exist $\tau' \in [0, \tau_f]$ such that $\hat{\mathbf{x}}(\tau') \in \theta_E$; and since $\hat{\mathbf{x}}(0) = x^i$ and $\hat{\mathbf{x}}(\tau_f) = x^f \in \theta_P$, the path $\hat{\pi}$ that emanates from x^i, is generated by $(p^*_{x^0}, e^*_{x^0})$ and is given by function $\hat{\mathbf{x}}$ reaches θ_P. That is in contradiction with Lemma 5.2, since $\hat{\pi}$ is an $\mathscr{E}$-path and hence must belong to S_E; that is, it cannot reach θ_P since $\theta_P \subseteq S_P$. Thus we have established that $\lambda^f \cdot \eta^f = 0$ for all η^f in (6.27).

Now let us assume that target θ_P is defined by $\boldsymbol{\theta}(x) \leqslant 0$.

We shall prove that

$$\lambda^f = K \operatorname{grad} \boldsymbol{\theta}(x^f), \qquad K = \text{constant} > 0 \tag{6.28$'$}$$

Suppose that $\lambda^f = K \operatorname{grad} \boldsymbol{\theta}(x^f)$ with $K < 0$, and let

$$\eta^f \triangleq - \operatorname{grad} \boldsymbol{\theta}(x^f)$$

Then

$$\lambda^f \cdot \eta^f > 0$$

Furthermore there exists $\alpha > 0$ such that for all ϵ, $0 < \epsilon < \alpha$,

$$x^f + \epsilon\eta^f \in \theta_P$$

If Assumption 6.2 is satisfied along π_{pe}, then for ϵ sufficiently small there exists a solution $\hat{\mathbf{x}}\colon\ \tau \to x = \hat{\mathbf{x}}(\tau)$, $\tau \in [0, \tau_f]$, of trajectory equation (6.2) with end condition

$$\hat{\mathbf{x}}(\tau_f) = x^f + \epsilon\eta^f$$

and there exists a point $x^i = \hat{\mathbf{x}}(0)$ such that

$$x^i = x^0 + \epsilon\eta^0 + o(\epsilon)$$

with $\eta^0 = \boldsymbol{\eta}(0)$, where $\boldsymbol{\eta}$ is the solution of variational equation (6.3) with end condition $\eta^f = \boldsymbol{\eta}(\tau_f)$.

We have

$$\mathbf{g}(x^i) = \mathbf{g}(x^0 + \epsilon\eta^0 + o(\epsilon)) = \epsilon \operatorname{grad} \mathbf{g}(x^0) \cdot \eta^0 + o(\epsilon)$$

where $\lim_{\epsilon \to 0} o(\epsilon)/\epsilon = 0$. For ϵ sufficiently small, the sign of the right-hand side is the same as the sign of $\epsilon \operatorname{grad} \mathbf{g}(x^0) \cdot \eta^0$, and hence $\mathbf{g}(x^i)$ has the same sign as $\epsilon\lambda^0 \cdot \eta^0$.

Since $\epsilon > 0$, since $\boldsymbol{\lambda}(\tau) \cdot \boldsymbol{\eta}(\tau) = \text{constant}$ along π_{pe} and since $\lambda^f \cdot \eta^f > 0$, it follows that

$$\mathbf{g}(x^i) > 0$$

for ϵ sufficiently small.

Hence, for ϵ sufficiently small,

$$x^i \in S_E \cap \Delta(x^0)$$

Then, by the same arguments as those employed above, one obtains a contradiction with the fact that

$$x^f + \epsilon\eta^f \in \theta_P \subseteq S_P$$

Finally, since $K \neq 0$, we conclude that $K > 0$.

If $x^f \in \theta_E$ and θ_E is defined by

$$\boldsymbol{\theta}(x) \geqslant 0$$

one can again prove in similar fashion that

$$\lambda^f = K \operatorname{grad} \boldsymbol{\theta}(x^f), \qquad K > 0$$

6.10 A MIN-MAX PRINCIPLE

Now let

$$\mathscr{H}(\lambda, x, u, v) \triangleq \lambda \cdot f(x, u, v)$$

From Lemma 6.3 and from (6.28) one can deduce

Theorem 6.1. *Let π_{pe} be the boundary path represented by* $\mathbf{x}$: $\tau \to x = \mathbf{x}(\tau)$, $\tau \in [0, \tau_s]$, $\tau_s \neq 0$. *If Assumptions* 6.1–6.3 *are satisfied, then there exists a nonzero continuous solution* $\boldsymbol{\lambda}$: $\tau \to \lambda = \boldsymbol{\lambda}(\tau)$, $\tau \in [0, \tau_s]$, *of adjoint equation* (6.4) *such that*

$$\text{(a)} \qquad \min_{u \in K_u(x)} \mathscr{H}(\lambda, x, u, e^*_{x^0}(x)) = \max_{v \in K_v(x)} \mathscr{H}(\lambda, x, p^*_{x^0}(x), v)$$
$$= \mathscr{H}(\lambda, x, p^*_{x^0}(x), e^*_{x^0}(x))$$

$$\text{(b)} \qquad \mathscr{H}(\lambda, x, p^*_{x^0}(x), e^*_{x^0}(x)) = 0$$

for all $\tau \in [0, \tau_s]$, where x^0 is the initial point of π_{pe}, and $(p^*_{x^0}, e^*_{x^0})$ is the strategy pair that generates π_{pe}.

Furthermore, if $\theta_P \cap \theta_E = \varnothing$, if π_{pe} reaches θ_P (or θ_E) at point x^f, if Assumptions 6.4 *and* 6.5 *are satisfied, and if θ_P (or θ_E) is defined by $\boldsymbol{\theta}_P(x) \leqslant 0$ (or $\boldsymbol{\theta}_E(x) \geqslant 0$), then*

$$\text{(c)} \qquad \lambda^f = K \operatorname{grad} \boldsymbol{\theta}(x^f), \qquad K > 0$$

where $\boldsymbol{\theta} = \boldsymbol{\theta}_P$ or $\boldsymbol{\theta}_E$.

If $x^s \triangleq \mathbf{x}(\tau_s)$ does not belong to $\theta_P \cup \theta_E$, then Theorem 6.1 is a trivial consequence of Lemma 6.3 together with the conclusions of Section 6.8.

If $x^s \in \theta_P \cup \theta_E$, we shall consider a value of τ, say $\tau' \in [0, \tau_s)$, and a point $x' = \mathbf{x}(\tau') \in \pi_{pe}$, at which relations (a) and (b) above hold; namely,

$$\min_{u \in K_u(x')} \boldsymbol{\lambda}(\tau') \cdot f(x', u, e^*_{x^0}(x')) = \max_{v \in K_v(x')} \boldsymbol{\lambda}(\tau') \cdot f(x', p^*_{x^0}(x'), v)$$

$$= \boldsymbol{\lambda}(\tau') \cdot f(x', p^*_{x^0}(x'), e^*_{x^0}(x'))$$

and

$$\boldsymbol{\lambda}(\tau') \cdot f(x', p^*_{x^0}(x'), e^*_{x^0}(x')) = 0$$

Since f is of class C^1 on $G \times U \times V$, and $p^*_{x^0}$ and $e^*_{x^0}$ are of class C^1 on a neighborhood of π_{pe}, and $\mathbf{x}$: $\tau' \to x' = \mathbf{x}(\tau')$ as well as $\boldsymbol{\lambda}$: $\tau' \to \lambda' = \boldsymbol{\lambda}(\tau')$ are continuous functions of τ' on $[0, \tau_s]$, it follows that H, defined by

$$H: \quad \tau' \to H(\tau')$$

where

$$H(\tau') = \boldsymbol{\lambda}(\tau') \cdot f(\mathbf{x}(\tau'), p^*_{x^0}(\mathbf{x}(\tau')), e^*_{x^0}(\tau')))$$

is a continuous function of τ' on $[0, \tau_s]$. Hence, relation (b) holds at $\tau = \tau_s$ and $x^s \triangleq \mathbf{x}(\tau_s)$.

It is clear that relation (a) also holds at $\tau = \tau_s$, since

$$\boldsymbol{\lambda}(\tau_s) \cdot f(\mathbf{x}(\tau_s), u^b, e^*_{x^0}(\mathbf{x}(\tau_s))) \geqslant 0$$

and

$$\boldsymbol{\lambda}(\tau_s) \cdot f(\mathbf{x}(\tau_s), p^*_{x^0}(\mathbf{x}(\tau_s), v^b) \leqslant 0$$

can be deduced from

$$\boldsymbol{\lambda}(\tau') \cdot f(\mathbf{x}(\tau'), \alpha_{x^s}(\mathbf{x}(\tau')), e^*_{x^0}(\mathbf{x}(\tau'))) \geqslant 0, \qquad \alpha_{x^s}(x^s) = u^b$$

and

$$\boldsymbol{\lambda}(\tau') \cdot f(\mathbf{x}(\tau'), p^*_{x^0}(\mathbf{x}(\tau')), \beta_{x^s}(\mathbf{x}(\tau'))) \leqslant 0, \qquad \beta_{x^s}(x^s) = v^b$$

by continuity, for every given $u^b \in K_u(x^s)$ and every given $v^b \in K_v(x^s)$.

6.11 AUTONOMOUS GAMES WITH TIME-INDEPENDENT TARGETS

An autonomous game is one in which the function f, in the right-hand side of the state equation (6.1), does not contain the variable $x_n = t$ explicitly. This situation can arise only if p and e do not contain x_n in their arguments; that is, if they are functions of $x_1, \ldots, x_{n-1}$ only.

Let $\boldsymbol{\gamma}(\tau) \triangleq \text{grad}\, \mathbf{g}(\mathbf{x}(\tau))$. From (6.23) we obtain†

$$\frac{d}{d\tau}(\boldsymbol{\gamma}(\tau)\cdot\boldsymbol{\eta}(\tau)) = \frac{d\boldsymbol{\gamma}(\tau)}{d\tau}\cdot\boldsymbol{\eta}(\tau) + \boldsymbol{\gamma}(\tau)\cdot\frac{d\boldsymbol{\eta}(\tau)}{d\tau} = 0$$

or, taking account of (6.3),

$$\frac{d\boldsymbol{\gamma}(\tau)}{d\tau}\cdot\boldsymbol{\eta}(\tau) = -\boldsymbol{\gamma}(\tau)\cdot\left(\frac{\partial f}{\partial x} + \frac{\partial f}{\partial u}\frac{\partial p_x^{*0}}{\partial x} + \frac{\partial f}{\partial v}\frac{\partial e_x^{*0}}{\partial x}\right)\Bigg|_{x=\mathbf{x}(\tau)}\boldsymbol{\eta}(\tau)$$

or equivalently,

$$\boldsymbol{\eta}(\tau)\cdot\frac{d\boldsymbol{\gamma}(\tau)}{d\tau} = -\boldsymbol{\eta}(\tau)\cdot\left(\frac{\partial f}{\partial x} + \frac{\partial f}{\partial u}\frac{\partial p_x^{*0}}{\partial x} + \frac{\partial f}{\partial v}\frac{\partial e_x^{*0}}{\partial x}\right)^{\mathrm{T}}\Bigg|_{x=\mathbf{x}(\tau)}\boldsymbol{\gamma}(\tau) \quad (6.30)$$

Now recall that $f_n(x, u, v) \equiv 1$, and so $\partial f_n/\partial x_\alpha = 0$ for $\alpha = 1, \ldots, n$. It follows that the last row of the matrix

$$\left(\frac{\partial f}{\partial x} + \frac{\partial f}{\partial u}\frac{\partial p_x^{*0}}{\partial x} + \frac{\partial f}{\partial v}\frac{\partial e_x^{*0}}{\partial x}\right)^{\mathrm{T}}$$

is zero. Moreover, since

$$\frac{\partial f_\nu}{\partial x_n} = \frac{\partial p_\nu}{\partial x_n} = \frac{\partial e_\nu}{\partial x_n} = 0‡$$

for $\nu = 1, \ldots, n$ in an autonomous game, it follows that the last column of this matrix is also zero.

On the other hand, if targets θ_P and θ_E are time-independent, that is, if θ_P and θ_E in two target problems, and θ_P or θ_E in single-target problems,§ are x_n-cylindrical sets in E^n, then the map of the game is composed of x_n-cylindrical sets in E^n. It follows that S_P, S_E and Ξ are x_n-cylindrical sets in E^n. In this case, $\mathbf{g}$ is a function of $x_1, \ldots, x_{n-1}$ only, which implies that the nth component of vector $\boldsymbol{\gamma}(\tau)$ vanishes; namely,

$$\boldsymbol{\gamma}(\tau) = (\gamma_1(\tau), \gamma_2(\tau), \ldots, \gamma_{n-1}(\tau), 0)$$

Since (6.30) is satisfied by $n - 1$ linearly independent vectors $\boldsymbol{\eta}(\tau)$ at point $x = \mathbf{x}(\tau)$, we arrive at

$$\frac{d\boldsymbol{\gamma}(\tau)}{d\tau} = -\left(\frac{\partial f}{\partial x} + \frac{\partial f}{\partial u}\frac{\partial p_x^{*0}}{\partial x} + \frac{\partial f}{\partial v}\frac{\partial e_x^{*0}}{\partial x}\right)^{\mathrm{T}}\Bigg|_{x=\mathbf{x}(\tau)}\boldsymbol{\gamma}(\tau)$$

† Here we require that $\mathbf{g}$ is twice continuously differentiable.

‡ Here, $p = p_x^{*0}$ and $e = e_x^{*0}$, and p_ν and e_ν are the νth components of vectors p and e, respectively.

§ Of course, Ξ may be empty; and, in one target games with $\theta_P \neq \emptyset$ and $\theta_E = \emptyset$, S_E may be empty.

That is, function $\boldsymbol{\gamma}$: $\tau \to \boldsymbol{\gamma}(\tau)$, $\tau \in [0, \tau_s]$, where

$$\boldsymbol{\gamma}(\tau) = \operatorname{grad} \mathbf{g}(\mathbf{x}(\tau))$$

is a solution of adjoint equation (6.4).

Finally, according to the definition of $\boldsymbol{\lambda}(\tau)$, we have†

$$\boldsymbol{\lambda}(\tau) = k \operatorname{grad} \mathbf{g}(\mathbf{x}(\tau)) \tag{6.31}$$

where k is a positive constant.

6.12 CONSTRAINTS

Let us consider constraint sets $K_u(x)$ and $K_v(x)$, respectively, given by

$$\begin{aligned} \varphi_\alpha(x, u) &\leqslant 0, \qquad \alpha = 1, 2, \ldots, k \\ \psi_\alpha(x, v) &\leqslant 0, \qquad \alpha = 1, 2, \ldots, l \end{aligned} \tag{6.32}$$

where functions φ_α and ψ_α are of class C^1 on $G \times U$ and $G \times V$, respectively.

At a point $x' = \mathbf{x}(\tau')$, $\tau' \in [0, \tau_s]$, of a boundary path π_{pe} emanating from $x^0 \in \Xi$, some or all of inequality constraints (6.32) may be equalities. Suppose that

$$\begin{aligned} \varphi_\alpha(x', p^*_{x^0}(x')) &= 0, \qquad \alpha = 1, 2, \ldots, k' \leqslant k \\ \psi_\alpha(x', e^*_{x^0}(x')) &= 0, \qquad \alpha = 1, 2, \ldots, l' \leqslant l \end{aligned}$$

We shall suppose also that if $k > r$ or $l > s$, at most r of the $\varphi_\alpha(x, u)$ or s of the $\psi_\alpha(x, u)$ vanish at any point of $G \times U$ and $G \times V$, respectively, and furthermore that the matrices

$$\left[\frac{\partial \varphi_\alpha(x, u)}{\partial u_\sigma}\right], \qquad \begin{aligned} \alpha &= 1, 2, \ldots, k' \\ \sigma &= 1, 2, \ldots, r \end{aligned}$$

$$\left[\frac{\partial \psi_\alpha(x, u)}{\partial v_\sigma}\right], \qquad \begin{aligned} \alpha &= 1, 2, \ldots, l' \\ \sigma &= 1, 2, \ldots, s \end{aligned}$$

have maximum rank at $u = p^*_{x^0}(x)$, $v = e^*_{x^0}(x)$, $x = \mathbf{x}^*(k)$.

By arguments similar to those employed in Chapter 3, one can verify that adjoint equation (6.4) may be written

$$\frac{d\lambda}{d\tau} = -\left(\frac{\partial f}{\partial x}\right)^{\mathrm{T}} \lambda - \left(\frac{\partial \varphi}{\partial x}\right)^{\mathrm{T}} \mu - \left(\frac{\partial \psi}{\partial x}\right)^{\mathrm{T}} \nu \tag{6.33}$$

† At all points $\mathbf{x}(\tau)$ where grad $\mathbf{g}(\mathbf{x}(\tau))$ is defined.

where

$$\frac{\partial \varphi}{\partial x} \triangleq \left[\frac{\partial \varphi_\nu(x, u)}{\partial x_\alpha}\right] \qquad \begin{matrix} \nu = 1, 2, \ldots, k \\ \alpha = 1, 2, \ldots, n \end{matrix}$$

$$\frac{\partial \psi}{\partial x} \triangleq \left[\frac{\partial \psi_\nu(x, v)}{\partial x_\alpha}\right] \qquad \begin{matrix} \nu = 1, 2, \ldots, l \\ \alpha = 1, 2, \ldots, n \end{matrix}$$

evaluated at $x = \mathbf{x}(\tau)$, $u = p^*_{x^0}(\mathbf{x}(\tau))$, $v = e^*_{x^0}(\mathbf{x}(\tau))$; and Lagrange multipliers μ and ν are k- and l- component vectors, respectively, given by

$$\mu = \begin{pmatrix} \mu_1 \\ \mu_2 \\ \cdot \\ \cdot \\ \cdot \\ \mu_{k'} \\ 0 \\ \cdot \\ \cdot \\ \cdot \\ 0 \end{pmatrix} \qquad \nu = \begin{pmatrix} \nu_1 \\ \nu_2 \\ \cdot \\ \cdot \\ \cdot \\ \nu_{l'} \\ 0 \\ \cdot \\ \cdot \\ \cdot \\ 0 \end{pmatrix} \text{with } \mu_i \geqslant 0,\ \nu_i \leqslant 0.$$

Thus, if constraints (6.32) are independent of x, adjoint equation (6.4) reduces to

$$d\lambda/d\tau = -(\partial f/\partial x)^{\mathrm{T}}\lambda \tag{6.34}$$

6.13 SEMIPERMEABLE SURFACES

We shall now complement Theorem 6.1 by

Theorem 6.2. *If* $\mathbf{g}\colon\ x \to \mathbf{g}(x)$, $x \in G$, *is such that* grad $\mathbf{g}(x)$ *is defined and is different from zero for all* $x \in G$, *and if there exist strategies* $\tilde{p}$ *and* $\tilde{e}$ *such that*

(a) $$\min_{u \in K_u(x)} \mathbf{n}(x) \cdot f(x, u, \tilde{e}(x)) = \max_{v \in K_v(x)} \mathbf{n}(x) \cdot f(x, \tilde{p}(x), v) = \mathbf{n}(x) \cdot f(x, \tilde{p}(x), \tilde{e}(x)) \tag{6.35}$$

and

(b) $$\mathbf{n}(x) \cdot f(x, \tilde{p}(x), \tilde{e}(x)) = 0 \tag{6.36}$$

for all $x \in G$, *where* $\mathbf{n}(x) \triangleq \operatorname{grad} \mathbf{g}(x)$; *and if there exists a constant* C *such that*

$$\mathbf{g}(x) < C \qquad \forall x \in \theta_P$$
$$[\text{or} \quad \mathbf{g}(x) > C \qquad \forall x \in \theta_E]$$

and if the surface $\mathscr{C}(C)$ *defined by*

$$\mathscr{C}(C) \triangleq \{x: \quad \mathbf{g}(x) = C\} \tag{6.37}$$

is not empty; then

$$\mathscr{C}(C) \subset \operatorname{comp} S_P$$
$$[\text{or} \quad \mathscr{C}(C) \subset \operatorname{comp} S_E]$$

Furthermore, if $\theta_E = \varnothing$ *and, of course,* $\theta_P \neq \varnothing$, *then* $\mathscr{C}(C) \subset S_E$.

Let $x^i \in \mathscr{C}(C)$ and consider describing curve $\Gamma_{\mathbf{x}}(x^i, p, \tilde{e})$, $\mathbf{x} \in \gamma(x^i, p, \tilde{e})$, where p is any strategy for J_P. If $\Gamma_{\mathbf{x}}(x^i, p, \tilde{e})$ is a single point, then obviously

$$\Gamma_{\mathbf{x}}(x^i, p, \tilde{e}) \cap \theta_P = \varnothing$$
$$[\text{or} \quad \Gamma_{\mathbf{x}}(x^i, p, \tilde{e}) \cap \theta_E = \varnothing]$$

Therefore let us suppose that $\Gamma_{\mathbf{x}}(x^i, p, \tilde{e})$ is given by $\mathbf{x}: \quad \tau \to x = \mathbf{x}(\tau)$, $\tau \in [0, \tau_s)$, where $\tau_s \neq 0$.

As a consequence of (6.35),

$$\operatorname{grad} \mathbf{g}(\mathbf{x}(\tau)) \cdot f(\mathbf{x}(\tau), p(\mathbf{x}(\tau)), \tilde{e}(\mathbf{x}(\tau))) \geqslant 0 \tag{6.38}$$

for all $\tau \in [0, \tau_s)$; and since the left-hand side of (6.38) is the time derivative of $\mathbf{g}(\mathbf{x}(\tau))$, namely $d\mathbf{g}(\mathbf{x}(\tau))/d\tau$, for all $\tau \in (0, \tau_s)$, it follows that

$$\mathbf{g}_\tau: \tau \to \mathbf{g}_\tau(\tau),$$

where $\mathbf{g}_\tau(\tau) \triangleq \mathbf{g}(\mathbf{x}(\tau))$, is a non-decreasing function of τ on $[0, \tau_s)$.

Since $x^i = \mathbf{x}(0)$ belongs to $\mathscr{C}(C)$,

$$\mathbf{g}(\mathbf{x}(0)) = C$$

and if $\mathbf{g}(x) < C$ for all $x \in \theta_P$, then

$$\Gamma_{\mathbf{x}}(x^i, p, \tilde{e}) \cap \theta_P = \varnothing \tag{6.39}$$

for all $\mathbf{x} \in \gamma(x^i, p, \tilde{e})$ and for all $p \in \mathscr{S}_P$.

Since (6.39) holds for all $x^i \in \mathscr{C}(C)$,

$$(p, \tilde{e}) \notin \mathscr{T}_P(x^i)$$
$$\forall p \in \mathscr{S}_P \qquad \forall x^i \in \mathscr{C}(C)$$

Consequently,

$$\mathscr{C}(C) \subset \text{comp } A$$

which implies that

$$\mathscr{C}(C) \subset \text{comp } S_P$$

By similar arguments, one can prove that

$$\mathscr{C}(C) \subset \text{comp } S_E$$

if $\mathbf{g}(x) > C$ for all $x \in \theta_E$.

Hence, Theorem 6.2 is established.

Now, if $\theta_E = \varnothing$ and, of course, $\theta_P \neq \varnothing$, Equation (6.39) implies that

$$(p, \tilde{e}) \in \mathscr{T}_E^*(x^i)$$

$$\forall p \in \mathscr{S}_P \qquad \forall x^i \in \mathscr{C}(C)$$

Hence

$$\mathscr{C}(C) \subset S_E$$

Definition 6.2. A surface $\mathscr{C}(C)$ defined by (6.37) is called a *semipermeable surface,*† if grad $\mathbf{g}(x)$ is defined and is different from zero for all $x \in G$, and if there exist strategies $\tilde{p}$ and $\tilde{e}$ such that Conditions (a) and (b) of Theorem 6.2 are satisfied.

We shall complement Theorem 6.2 by

Theorem 6.3. *If* $\mathbf{g}_\nu$: $x \to \mathbf{g}_\nu(x)$, $\nu \in I$, *where* I *is a set of indices, are functions defined on* G, *such that* grad $\mathbf{g}_\nu(x)$ *is defined and is different from zero for all* $x \in G$ *and for all* $\nu \in I$, *and if there exists a strategy* $\tilde{e}$ *such that*

$$\mathbf{n}_\nu(x) \cdot f(x, u, \tilde{e}(x)) \geqslant 0$$

for all $u \in K_u(x)$, $x \in G$, $\nu \in I$, *where* $\mathbf{n}_\nu(x) \triangleq$ grad $\mathbf{g}_\nu(x)$; *and if there exist constants* C_ν, $\nu \in I$, *such that*

$$\mathscr{D}_E \neq \varnothing \qquad \text{and} \qquad \theta_P \cap \mathscr{D}_E = \varnothing$$

where

$$\mathscr{D}_E \triangleq \{x: \quad \mathbf{g}_\nu(x) \geqslant C_\nu, \quad \nu \in I\}$$

then

$$\mathscr{D}_E \subset \text{comp } S_P$$

Furthermore, if $\theta_E = \varnothing$ *and, of course,* $\theta_P \neq \varnothing$, *then*

$$\mathscr{D}_E \subset S_E$$

† According to the terminology of Isaacs.

Let $x^i \in \mathscr{D}_E$ and consider describing curve $\Gamma_{\mathbf{x}}(x^i, p, \tilde{e})$, $\mathbf{x} \in \gamma(x^i, p, \tilde{e})$, where p is any strategy for J_P. If $\Gamma_{\mathbf{x}}(x^i, p, \tilde{e})$ is a single point, then, obviously, $\Gamma_{\mathbf{x}}(x^i, p, \tilde{e}) \cap \theta_P = \varnothing$.

So let us suppose that $\Gamma_{\mathbf{x}}(x^i, p, \tilde{e})$ is given by $\mathbf{x}$: $\tau \to x = \mathbf{x}(\tau)$, $\tau \in [0, \tau_s)$, where $\tau_s \neq 0$.

We have

$$\operatorname{grad} \mathbf{g}_\nu(\mathbf{x}(\tau)) \cdot f(\mathbf{x}(\tau), p(\mathbf{x}(\tau)), \tilde{e}(\mathbf{x}(\tau))) \geqslant 0 \qquad \nu \in I$$

for all $\tau \in [0, \tau_s)$; and hence

$$d\mathbf{g}_\nu(\mathbf{x}(\tau))/d\tau \geqslant 0 \qquad \nu \in I$$

for all $\tau \in (0, \tau_s)$, from which it follows immediately that

$$\Gamma_{\mathbf{x}}(x^i, p, \tilde{e}) \subset \mathscr{D}_E$$

and hence

$$\Gamma_{\mathbf{x}}(x^i, p, \tilde{e}) \cap \theta_P = \varnothing$$

for all $\mathbf{x} \in \gamma(x^i, p, \tilde{e})$, for all $p \in \mathscr{S}_P$ and for all $x^i \in \mathscr{D}_E$. Consequently,

$$(p, \tilde{e}) \notin \mathscr{T}_P{}^*(x^i) \qquad \forall p \in \mathscr{S}_P \qquad \forall x^i \in \mathscr{D}_E$$

and so

$$\mathscr{D}_E \subset \operatorname{comp} S_P$$

Now, if $\theta_E = \varnothing$ and, of course, $\theta_P \neq \varnothing$, then

$$(p, \tilde{e}) \in \mathscr{T}_E{}^*(x^i)$$

$$\forall p \in \mathscr{S}_P \qquad \forall x^i \in \mathscr{D}_E$$

Hence

$$\mathscr{D}_E \subset S_E$$

which proves Theorem 6.3.

By similar arguments as in the proof of Theorem 6.3 one can establish

Theorem 6.4. *If* $\mathbf{g}_\nu$: $x \to \mathbf{g}_\nu(x)$, $\nu \in I$, *where* I *is a set of indices, are functions defined on* G, *such that* grad $\mathbf{g}_\nu(x)$ *is defined and is different from zero for all* $x \in G$ *and for all* $\nu \in I$, *and if there exists a strategy* $\tilde{p}$ *such that*

$$\mathbf{n}_\nu(x) \cdot f(x, \tilde{p}(x), v) \leqslant 0$$

for all $v \in K_v(x)$, $x \in G$, $\nu \in I$, *where* $\mathbf{n}_\nu(x) \triangleq \operatorname{grad} \mathbf{g}_\nu(x)$; *and if there exist constants* C_ν, $\nu \in I$, *such that* $\mathscr{D}_P \neq \varnothing$ *and* $\theta_P \cap \mathscr{D}_P = \varnothing$, *where*

$$\mathscr{D}_P \triangleq \{x: \quad \mathbf{g}_\nu(x) \leqslant C_\nu, \nu \in I\}$$

then

$$\mathscr{D}_P \subset \text{comp } S_E$$

Furthermore, if $\theta_P = \varnothing$ *and, of course* $\theta_E \neq \varnothing$, *then*

$$\mathscr{D}_P \subset S_P$$

6.14 A PROPERTY OF THE GRADIENT ALONG A PATH IN A SEMIPERMEABLE SURFACE

Consider a non-null path $\tilde{\pi}$ emanating from $x^i \in \mathscr{C}(C)$, generated by strategy pair $(\tilde{p}, \tilde{e})$ and represented by $\mathbf{x}\colon \quad \tau \to x = \mathbf{x}(\tau), \quad \tau \in [0, \tau_j]$. As a consequence of (6.36)

$$\frac{d\mathbf{g}(\mathbf{x}(\tau))}{d\tau} = \text{grad } \mathbf{g}(\mathbf{x}(\tau)) \cdot f(\mathbf{x}(\tau), \tilde{p}(\mathbf{x}(\tau)), \tilde{e}(\mathbf{x}(\tau))) = 0 \qquad (6.40)$$

and so $\mathbf{g}(\mathbf{x}(\tau)) = C$ for all $\tau \in [0, \tau_j]$. Hence, $\tilde{\pi} \subset \mathscr{C}(C)$.

Now, we introduce

Assumption 6.6. Strategies $\tilde{p}$ and $\tilde{e}$ are of class C^1 on a neighborhood of $\tilde{\pi}$.

Relation (6.36) is an identity in x on G. Let us rewrite it as

$$\sum_{\alpha=1}^{n} \frac{\partial \mathbf{g}(x)}{\partial x_\alpha} f_\alpha(x, \tilde{p}(x), \tilde{e}(x)) = 0 \qquad (6.41)$$

By differentiating (6.41) with respect to x_β, $\beta = 1, 2, \ldots, n$, we obtain†

$$\sum_{\alpha=1}^{n} \frac{\partial^2 \mathbf{g}(x)}{\partial x_\alpha \, \partial x_\beta} f_\alpha(x, \tilde{p}(x), \tilde{e}(x)) + \sum_{\alpha=1}^{n} \frac{\partial \mathbf{g}(x)}{\partial x_\alpha} \frac{\partial f_\alpha(x, \tilde{p}(x), \tilde{e}(x))}{\partial x_\beta} = 0, \qquad \beta = 1, 2, \ldots, n \quad (6.42)$$

Now, along a non-null path $\tilde{\pi}$ lying on $\mathscr{C}(C)$,

$$\frac{d}{d\tau}\left(\frac{\partial \mathbf{g}(x)}{\partial x_\beta}\right) = \sum_{\alpha=1}^{n} \frac{\partial^2 \mathbf{g}(x)}{\partial x_\alpha \, \partial x_\beta} f_\alpha(x, \tilde{p}(x), \tilde{e}(x)), \qquad \beta = 1, 2, \ldots, n \quad (6.43)$$

† Here we require that $\mathbf{g}$ is twice continuously differentiable.

Equations (6.42) and (6.43) lead to

$$\frac{d}{d\tau}\left(\frac{\partial \mathbf{g}(x)}{\partial x_\beta}\right) = -\sum_{\alpha=1}^{n} \frac{\partial \mathbf{g}(x)}{\partial x_\alpha} \frac{\partial f_\alpha(x, \tilde{p}(x), \tilde{e}(x))}{\partial x_\beta}, \qquad \beta = 1, 2, \ldots, n \quad (6.44)$$

By developing $\partial f_\alpha(x, \tilde{p}(x), \tilde{e}(x))/\partial x_\beta$ and writing (6.44) in vector form, we obtain

$$\frac{d}{d\tau}\mathbf{n}(x) = -\left(\frac{\partial f}{\partial x} + \frac{\partial f}{\partial u}\frac{\partial \tilde{p}}{\partial x} + \frac{\partial f}{\partial v}\frac{\partial \tilde{e}}{\partial x}\right)^{\mathrm{T}} \mathbf{n}(x) \tag{6.45}$$

for $x = \mathbf{x}(\tau) \in \tilde{\pi}$

If constraint sets $K_u(x)$ and $K_v(x)$ are given by (6.32), one can verify, by arguments similar to those used in Chapter 3, that adjoint equation (6.45) may be written as

$$\frac{d\lambda}{d\tau} = -\left(\frac{\partial f}{\partial x}\right)^{\mathrm{T}} \lambda - \left(\frac{\partial \varphi}{\partial x}\right)^{\mathrm{T}} \mu - \left(\frac{\partial \psi}{\partial x}\right)^{\mathrm{T}} \nu \tag{6.46}$$

where $\lambda = \mathbf{n}(\mathbf{x}(\tau))$, and μ and ν have the same meaning as in Sec. 6.12.

If constraints (6.32) are independent of x, adjoint equation (6.45) reduces to

$$\frac{d\lambda}{d\tau} = -\left(\frac{\partial f}{\partial x}\right)^{\mathrm{T}} \lambda \tag{6.47}$$

Clearly, there are similarities between the properties of semipermeable surfaces and the properties of game surface Ξ. One can easily find examples in which Ξ is imbedded in a family of semipermeable surfaces.

6.15 TRANSVERSALITY CONDITION FOR A SEMIPERMEABLE SURFACE

Let $\mathbf{g}$: $x \to \mathbf{g}(x)$ be a function defined on G, such that grad $\mathbf{g}(x)$ is defined and is different from zero for all $x \in G$, and let us suppose that there exist strategies $\tilde{p}$ and $\tilde{e}$ such that Conditions (a) and (b) of Theorem 6.2 are satisfied.

Furthermore, let us suppose that $\max_{x \in \theta_P} \mathbf{g}(x)$ is defined, and let $M \triangleq \max_{x \in \theta_P} \mathbf{g}(x) \cdot \mathscr{C}(M) \triangleq \{x\colon \; \mathbf{g}(x) = M\}$ is a semipermeable surface. Clearly, we have

$$\mathscr{C}(M) \cap \theta_P \neq \varnothing$$

Let $\hat{x}_P \in \mathscr{C}(M) \cap \theta_P$. Since $\mathbf{g}(\hat{x}_P) = M$ and grad $\mathbf{g}(\hat{x}_P) \neq 0$, $\hat{x}_P$ is not an interior point of θ_P; that is, $\hat{x}_P \in \partial\theta_P$.

Let

$$\hat{x}_P + \epsilon\eta + o(\epsilon) \in \partial\theta_P$$

where $\hat{x}_P + \eta$ belongs to the tangent plane $T(\hat{x}_P)$ of $\partial\theta_P$ through $\hat{x}_P$; and $\lim_{\epsilon\to 0} \|o(\epsilon)\|/\epsilon = 0$.

Since $\hat{x}_P + \epsilon\eta + 0(\epsilon) \in \theta_P$, we have

$$\mathbf{g}(\hat{x}_P + \epsilon\eta + o(\epsilon)) = \mathbf{g}(\hat{x}_P) + \epsilon\eta \cdot \text{grad}\, \mathbf{g}(\hat{x}_P) + o(\|\epsilon\eta + o(\epsilon)\|) \leqslant M$$

and since $\mathbf{g}(\hat{x}_P) = M$, we have

$$\epsilon\eta \cdot \text{grad}\, \mathbf{g}(\hat{x}_P) + o(\|\epsilon\eta + o(\epsilon)\|) \leqslant 0$$

Dividing by ϵ, and allowing $\epsilon \to 0$, we obtain

(i) for $\epsilon > 0$

$$\eta \cdot \text{grad}\, \mathbf{g}(\hat{x}_P) \leqslant 0$$

and

(ii) for $\epsilon < 0$

$$\eta \cdot \text{grad}\, \mathbf{g}(\hat{x}_P) \geqslant 0$$

Hence

$$\eta \cdot \text{grad}\, \mathbf{g}(\hat{x}_P) = 0$$

Since this equation holds for all $\hat{x}_P + \eta \in T(\hat{x}_P)$ it follows that grad $\mathbf{g}(\hat{x}_P)$ is perpendicular to $T(\hat{x}_P)$. Hence, $\partial\theta_P$ and $\mathscr{C}(M)$ have the same tangent plane at point $\hat{x}_P$. It follows that

$$\text{grad}\, \mathbf{g}(\hat{x}_P) = K_P\, \text{grad}\, \boldsymbol{\theta}_P(\hat{x}_P)$$

Furthermore, if target θ_P is defined by $\boldsymbol{\theta}_P(x) \leqslant 0$, constant K_P is positive. By similar arguments one can prove that, if $\min_{x\in\theta_E} \mathbf{g}(x)$ is defined, then

$$\text{grad}\, \mathbf{g}(\hat{x}_E) = K_E\, \text{grad}\, \boldsymbol{\theta}_E(\hat{x}_E)$$

where

$$\hat{x}_E \in \mathscr{C}(m) \cap \theta_E$$

$$\mathscr{C}(m) \triangleq \{x: \;\; \mathbf{g}(x) = m\}$$

$$m \triangleq \min_{x\in\theta_E} \mathbf{g}(x)$$

$$K_E > 0 \text{ if target } \theta_E \text{ is defined by } \boldsymbol{\theta}_E(x) \geqslant 0$$

Surface $\mathscr{C}(M)$ separates G into two disjoint sets, namely

$$\mathscr{E} \triangleq \{x: \quad \mathbf{g}(x) > M\}$$

and

$$\mathscr{E}' \triangleq \{x: \quad \mathbf{g}(x) < M\}$$

Likewise $\mathscr{C}(m)$ separates G into two disjoint sets, namely

$$\mathscr{P} \triangleq \{x: \quad \mathbf{g}(x) < m\}$$

and

$$\mathscr{P}' \triangleq \{x: \quad \mathbf{g}(x) > m\}$$

And, from Theorems 6.3 and 6.4

$$\mathscr{E} \subset \text{comp } S_P \qquad \text{and} \qquad \mathscr{P} \subset \text{comp } S_E.$$

Furthermore

(i) if $\theta_E = \varnothing$ and, indeed $\theta_P \neq \varnothing$, then $\mathscr{E} \subset S_E$; and
(ii) if $\theta_P = \varnothing$ and, indeed $\theta_E \neq \varnothing$, then $\mathscr{P} \subset S_P$.

6.16 A MIN–MAX PRINCIPLE ALONG A PATH IN A SEMI-PERMEABLE SURFACE

From the discussion of Sections 6.14–6.15 one can deduce at once

Theorem 6.5. *If $\tilde{\pi}$ is a non-null path emanating from a point x^i of a semi-permeable surface $\mathscr{C}(C) \triangleq \{x: \ \mathbf{g}(x) = C\}$, generated by strategy pair $(\tilde{p}, \tilde{e})$ and represented by $\mathbf{x}: \ \tau \to x = \mathbf{x}(\tau)$, $\tau \in [0, \tau_s]$, and if Assumption 6.6 is satisfied, then there exists a nonzero continuous solution $\boldsymbol{\lambda}: \ \tau \to \boldsymbol{\lambda}(\tau)$, $\tau \in [0, \tau_s]$, of adjoint equation*

$$\frac{d\lambda}{d\tau} = -\left(\frac{\partial f}{\partial x} + \frac{\partial f}{\partial u}\frac{\partial \tilde{p}}{\partial x} + \frac{\partial f}{\partial v}\frac{\partial \tilde{e}}{\partial x}\right)^{\mathrm{T}} \lambda$$

such that

(a)
$$\begin{aligned}\min_{u \in K_u(x)} &\mathscr{H}(\boldsymbol{\lambda}(\tau), \mathbf{x}(\tau), u, \tilde{e}(\mathbf{x}(\tau)))\\ &= \max_{v \in K_v(x)} \mathscr{H}(\boldsymbol{\lambda}(\tau), \mathbf{x}(\tau), \tilde{p}(\mathbf{x}(\tau)), v)\\ &= \mathscr{H}(\boldsymbol{\lambda}(\tau), \mathbf{x}(\tau), \tilde{p}(\mathbf{x}(\tau)), \tilde{e}(\mathbf{x}(\tau)))\end{aligned}$$

(b) $\mathscr{H}(\boldsymbol{\lambda}(\tau), \mathbf{x}(\tau), \tilde{p}(\mathbf{x}(\tau)), \tilde{e}(\mathbf{x}(\tau))) = 0$
for all $\tau \in [0, \tau_s]$.

Furthermore if Assumption 6.5 *is satisfied and if* θ_P *is defined by* $\boldsymbol{\theta}_P(x) \leqslant 0$ *and*

(i) $M \triangleq \max_{x \in \theta_P} \mathbf{g}(x)$ is defined;

(ii) $x^i \in \mathscr{C}(M)$;

(iii) π *reaches* θ_P *at point* x_P *at* τ_f^P,

then

(c) $\boldsymbol{\lambda}(\tau_f^P) = K_P \operatorname{grad} \boldsymbol{\theta}_P(x_P), \quad K_P > 0.$

Likewise, if Assumption 6.5 *is satisfied, and if* θ_E *is defined by* $\boldsymbol{\theta}_E(x) \geqslant 0$, *and*

(i)′ $m = \min_{x \in \theta_E} \mathbf{g}(x)$ *is defined*;

(ii)′ $x^i \in \mathscr{C}(m)$;

(iii)′ $\tilde{\pi}$ *reaches* θ_E *at point* x_E *at* τ_f^E,

then

(d) $\boldsymbol{\lambda}(\tau_f^E) = K_E \operatorname{grad} \boldsymbol{\theta}_E(x_E), \quad K_E > 0.$

6.17 EXAMPLE 6.1

Let us consider a one target qualitative game described by

$$\begin{aligned} dx_1/d\tau &= u + v \\ dx_2/d\tau &= -2 \\ dx_3/d\tau &= 1 \end{aligned} \tag{6.48}$$

with constraints given by

$$|u| \leqslant 1 \qquad |v| \leqslant 1 \tag{6.49}$$

and

$$\theta_P \triangleq \{x = (x_1, x_2, x_3)\colon \ \boldsymbol{\theta}(x) \triangleq x_1^2 + x_2^2 - R^2 \leqslant 0\} \tag{6.50}$$

Let us suppose that there exists a non-null boundary path π_{pe}, emanating from $x^0 \in \Xi$, generated by strategy pair $(p^*_{x^0}, e^*_{x^0})$, and represented by $\mathbf{x}\colon \ \tau \to x = \mathbf{x}(\tau)$, $\tau \in [0, \tau_f]$, which reaches θ_P at point $\mathbf{x}(\tau_f)$, and that Assumptions 6.2–6.4 are satisfied.†

The $\mathscr{H}$-function is given by

$$\mathscr{H}(\lambda, x, u, v) = \lambda_1(u + v) - 2\lambda_2 + \lambda_3 \tag{6.51}$$

† Assumption 6.1 is satisfied since the constraints are given by (6.49), and Assumption 6.5 is also satisfied since target θ_P is given by (6.50).

The adjoint equations are

$$d\lambda_\alpha/d\tau = 0, \qquad \alpha = 1, 2, 3 \tag{6.52}$$

so that

$$\boldsymbol{\lambda}_\alpha(\tau) = \text{constant}, \qquad \alpha = 1, 2, 3 \tag{6.53}$$

From condition (c) of Theorem 6.1 we have

$$\begin{aligned} \boldsymbol{\lambda}_1(\tau_f) &= c\mathbf{x}_1(\tau_f) \\ \boldsymbol{\lambda}_2(\tau_f) &= c\mathbf{x}_2(\tau_f) \qquad c > 0 \\ \boldsymbol{\lambda}_3(\tau_f) &= 0 \end{aligned} \tag{6.54}$$

where, in view of (6.50),

$$c^2 = R^{-2}[\boldsymbol{\lambda}_1^2(\tau_f) + \boldsymbol{\lambda}_2^2(\tau_f)] \tag{6.55}$$

Then it follows from (6.53) and (6.54) that

$$\begin{aligned} \boldsymbol{\lambda}_1(\tau) &\equiv c\mathbf{x}_1(\tau_f) \\ \boldsymbol{\lambda}_2(\tau) &\equiv c\mathbf{x}_2(\tau_f) \\ \boldsymbol{\lambda}_3(\tau) &\equiv 0 \end{aligned} \tag{6.56}$$

In view of condition (a) of Theorem 6.1,

$$\begin{aligned} \mathbf{u}(\tau) &= -\operatorname{sgn} \mathbf{x}_1(\tau_f) \\ \mathbf{v}(\tau) &= \operatorname{sgn} \mathbf{x}_1(\tau_f) \end{aligned} \tag{6.57}$$

and so by condition (b),

$$\boldsymbol{\lambda}_2(\tau) \equiv 0 \tag{6.58}$$

which implies that

$$\mathbf{x}_2(\tau_f) = 0$$

Along π_{pe}, (6.48) may be rewritten as

$$\begin{aligned} dx_1/d\tau &= 0 \\ dx_2/d\tau &= -2 \end{aligned} \tag{6.59}$$

Upon integration of (6.59) we find that a non-null boundary path π_{pe} that reaches θ_P at point $\mathbf{x}(\tau_f)$ is a straight line whose projection on the x_1–x_2 plane is given by

$$\begin{aligned} \mathbf{x}_1(\tau) &\equiv \mathbf{x}_1(\tau_f) = \pm R \\ \mathbf{x}_2(\tau) &= -2(\tau - \tau_f) \end{aligned} \tag{6.60}$$

The projection on the x_1–x_2 plane of the two families of possible boundary paths thus obtained is sketched on Fig. 6.6. One can prove that the closed shaded area of Fig. 6.6 is the projection of a set of E^3 that contains S_P, whereas the open unshaded area is the projection of a set of E^3 contained in S_E. Indeed, S_P and S_E depend on $\mathscr{S}_P$ and $\mathscr{S}_E$.

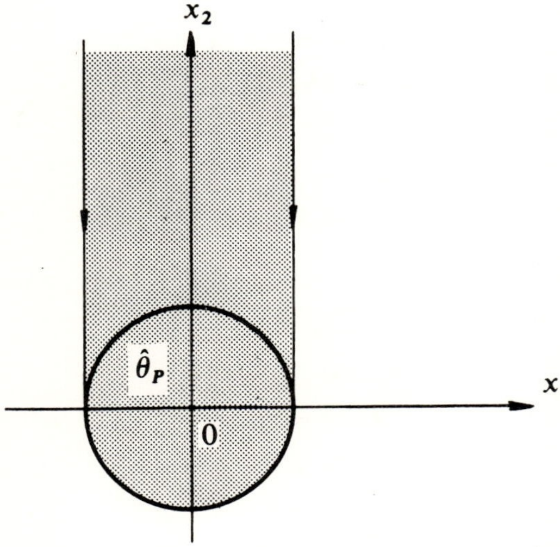

FIG. 6.6. Example 6.1. Projection on the x_1–x_2 plane of the two families of possible boundary paths; projection of a domain of E^3 that contains set S_P. ($\hat{\theta}_P$: projection of target θ_P.)

6.18 EXAMPLE 6.2

Next let us consider a one target qualitative game described by Eqs. (3.78), with constraints given by (3.79), and target θ_P defined by

$$\theta_P \triangleq \{x = (x_1, x_2, x_3){:}\quad \boldsymbol{\theta}(x) \triangleq {x_1}^2 + {x_2}^2 - R^2 \leqslant 0\} \tag{6.61}$$

Again let us suppose that there exists a non-null boundary path π_{pe} emanating from $x^0 \in \Xi$, generated by strategy pair $(p^*_{x^0}, e^*_{x^0})$, where $p^*_{x^0} = (p^*_{x^0 1}, p^*_{x^0 2})$, and represented by $\mathbf{x}{:}\quad \tau \to x = \mathbf{x}(\tau),\ \tau \in [0, \tau_f]$, which reaches θ_P at the point $\mathbf{x}(\tau_f)$, and that Assumptions 6.2–6.4 are satisfied.†

† Assumption 6.1 is satisfied since the constraints are given by (3.79), and Assumption 6.5 is also satisfied since target θ_P is given by (6.61).

The $\mathscr{H}$-function is given by

$$\mathscr{H}(\lambda, x, u, v) = \lambda_1 u_1 + \lambda_2 u_2 + \lambda_1 v + \lambda_3 \tag{6.62}$$

The adjoint equations are

$$d\lambda_\alpha/d\tau = 0, \qquad \alpha = 1, 2, 3 \tag{6.63}$$

so that

$$\boldsymbol{\lambda}_\alpha(\tau) = \text{constant}, \qquad \alpha = 1, 2, 3 \tag{6.64}$$

From condition (c) of Theorem 6.1 we have

$$\begin{aligned} \boldsymbol{\lambda}_1(\tau_f) &= c\mathbf{x}_1(\tau_f) \\ \boldsymbol{\lambda}_2(\tau_f) &= c\mathbf{x}_2(\tau_f) \qquad c > 0 \\ \boldsymbol{\lambda}_3(\tau_f) &= 0 \end{aligned} \tag{6.65}$$

where, in view of (6.61),

$$c^2 = R^{-2}[\boldsymbol{\lambda}_1^2(\tau_f) + \boldsymbol{\lambda}_2^2(\tau_f)] \tag{6.66}$$

Then it follows from (6.64) and (6.65) that

$$\begin{aligned} \boldsymbol{\lambda}_1(\tau) &\equiv c\mathbf{x}_1(\tau_f) \\ \boldsymbol{\lambda}_2(\tau) &\equiv c\mathbf{x}_2(\tau_f) \\ \boldsymbol{\lambda}_3(\tau) &\equiv 0 \end{aligned} \tag{6.67}$$

From condition (a) of Theorem 6.1 we obtain

$$\begin{aligned} \mathbf{u}_1(\tau) &= -w \frac{\boldsymbol{\lambda}_1(\tau)}{\sqrt{\boldsymbol{\lambda}_1^2(\tau) + \boldsymbol{\lambda}_2^2(\tau)}} = -w \frac{\mathbf{x}_1(\tau_f)}{R} \\ \mathbf{u}_2(\tau) &= -w \frac{\boldsymbol{\lambda}_2(\tau)}{\sqrt{\boldsymbol{\lambda}_1^2(\tau) + \boldsymbol{\lambda}_2^2(\tau)}} = -w \frac{\mathbf{x}_2(\tau_f)}{R} \\ \mathbf{v}(\tau) &= \operatorname{sgn} \boldsymbol{\lambda}_1(\tau) = \operatorname{sgn} \mathbf{x}_1(\tau_f) \end{aligned} \tag{6.68}$$

where

$$\mathbf{u}_1(\tau) = p^*_{x^0 1}(\mathbf{x}(\tau)), \qquad \mathbf{u}_2(\tau) = p^*_{x^0 2}(\mathbf{x}(\tau)), \qquad \mathbf{v}(\tau) = e^*_{x^0}(\mathbf{x}(\tau))$$

and so condition (b) leads to

$$-w\sqrt{\boldsymbol{\lambda}_1^2(\tau) + \boldsymbol{\lambda}_2^2(\tau)} + |\boldsymbol{\lambda}_1(\tau)| = 0 \tag{6.69}$$

Relations (6.64), (6.65), and (6.69) result in

$$\frac{|\boldsymbol{\lambda}_1(\tau_f)|}{\sqrt{\boldsymbol{\lambda}_1^2(\tau_f) + \boldsymbol{\lambda}_2^2(\tau_f)}} = \frac{|\mathbf{x}_1(\tau_f)|}{R} = w \tag{6.70}$$

Consequently,

$$|\mathbf{x}_1(\tau_f)| = Rw \tag{6.71}$$

at a point where a non-null boundary path π_{pe} reaches θ_P.

Along π_{pe}, (3.78) may be rewritten as

$$\begin{aligned} \frac{dx_1}{d\tau} &= -w\frac{\mathbf{x}_1(\tau_f)}{R} + \operatorname{sgn} \mathbf{x}_1(\tau_f) \\ \frac{dx_2}{d\tau} &= -w\frac{\mathbf{x}_2(\tau_f)}{R} \\ \frac{dx_3}{d\tau} &= 1 \end{aligned} \tag{6.72}$$

Upon integration of (6.72) we find that a non-null boundary path π_{pe} that reaches θ_P at point $\mathbf{x}(\tau_f)$ is a straight line whose projection on the x_1–x_2 plane is given by

$$x_2 = ax_1 + b \tag{6.73}$$

where

$$a \triangleq \frac{\mathbf{x}_2(\tau_f)}{\mathbf{x}_1(\tau_f) - (R/w) \operatorname{sgn} \mathbf{x}_1(\tau_f)} = \pm \frac{w}{\sqrt{1 - w^2}}$$

$$b \triangleq -(R/w)a \operatorname{sgn} \mathbf{x}_1(\tau_f) = \pm \frac{R}{\sqrt{1 - w^2}}$$

A straight line given by Eq. (6.73) intersects the x_1-axis at point P: $x_1 = R/w$, or at point Q: $x_1 = -R/w$, depending on the sign of $\mathbf{x}_1(\tau_f)$. Of course, it is tangent to the circle given by $x_1^2 + x_2^2 = R^2$.

The arguments above yield four families of possible boundary paths whose projections on the x_1–x_2 plane belong to the four half-rays $P'M'$, $P''M''$, $Q'N'$, $Q''N''$ sketched on Fig. 6.7.

Let us consider the case for which $a < 0$ and $b > 0$, namely

$$a = -\frac{w}{\sqrt{1 - w^2}}, \qquad b = \frac{R}{\sqrt{1 - w^2}}$$

In that case

$$\mathbf{x}_1(\tau_f) = Rw$$

$$\mathbf{x}_2(\tau_f) = R\sqrt{1 - w^2}$$

and hence,

$$\boldsymbol{\lambda}_1(\tau) \equiv cRw,$$

$$\boldsymbol{\lambda}_2(\tau) \equiv cR\sqrt{1 - w^2}$$

$$\boldsymbol{\lambda}_3(\tau) \equiv 0$$

Consider a plane $\mathscr{C}_1(C)$ given by

$$\mathbf{g}(x) \triangleq x_2 - ax_1 - b = C \qquad \text{with} \qquad C > 0$$

We have

$$\mathbf{g}(x) < C \qquad \forall x \in \theta_P$$

and

$$\mathbf{n}(x) \triangleq \text{grad}\,\mathbf{g}(x) = (-a, 1, 0) = [(Rc\sqrt{1 - w^2})^{-1}]\boldsymbol{\lambda}(\tau)$$

Conditions (a) and (b) of Theorem 6.2 are satisfied with

$$\tilde{p}(x) = \mathbf{u}(\tau) = (-w^2, -w\sqrt{1 - w^2})$$

$$\tilde{e}(x) = \mathbf{v}(\tau) = 1$$

and hence, according to Theorem 6.2, $\mathscr{C}_1(C)$ is a semipermeable surface such that

$$\mathscr{C}_1(C) \subset \text{comp}\, S_P \qquad \text{for all} \qquad C > 0$$

By similar arguments one can see that planes $\mathscr{C}_2(C)$, $\mathscr{C}_3(C)$, and $\mathscr{C}_4(C)$ given by

$$x_2 - ax_1 - b = C \qquad a > 0, b < 0, C < 0$$

$$x_2 - ax_1 - b = C \qquad a < 0, b < 0, C < 0$$

$$x_2 - ax_1 - b = C \qquad a > 0, b > 0, C > 0$$

respectively, are semipermeable surfaces such that

$$\left.\begin{array}{l}\mathscr{C}_2(C) \in \text{comp}\, S_P \\ \mathscr{C}_3(C) \in \text{comp}\, S_P\end{array}\right\} \qquad \text{for all} \qquad C < 0$$

$$\mathscr{C}_4(C) \in \text{comp}\, S_P \qquad \text{for all} \qquad C > 0$$

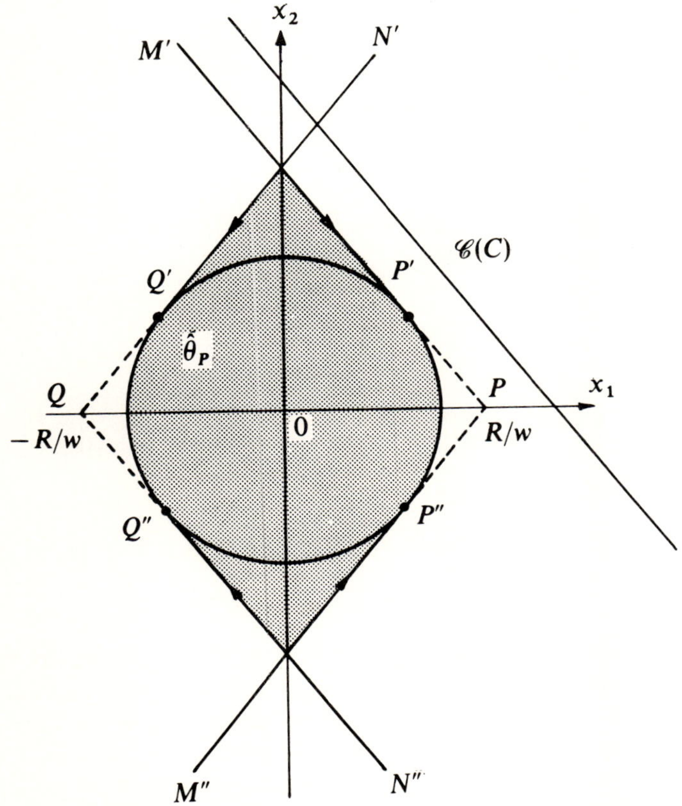

FIG. 6.7. Example 6.2. Projection on the x_1–x_2 plane of a domain of E^3 that contains set S_P. ($\hat{\theta}_P$: projection of target θ_P).

One can prove that the closed shaded area of Fig. 6.7 is the projection of a set of E^3 that contains S_P, whereas the open unshaded area is the projection of a set of E^3 contained in S_E. Indeed, S_P and S_E depend on $\mathscr{S}_P$ and $\mathscr{S}_E$.

We recommend that the reader compare the treatment of this example to the one of Example 3.1.

7 *A Connection between Qualitative and Quantitative Games*

7.1 PROBLEM STATEMENT, DEFINITIONS OF GAMES 1, 2, AND 3

Definition 7.1. As in Chapter 6, let us consider a qualitative game in E^n with state equation

$$dx/d\tau = f(x, u, v) \tag{7.1}$$

where $x \triangleq (x_1, x_2, \ldots, x_n)$. The definition of the game, the notation and the assumptions will be the same as those in Chapter 6, except for the fact that we shall now focus our attention on the case where $\theta_P \neq \varnothing$ and $\theta_E = \varnothing$. Such a game will be termed *Game 1.*

Now let us consider a new variable x_0 and the equation

$$dx_0/d\tau = f_0(x, u, v) \tag{7.2}$$

where f_0 is of class C^1 on $G \times U \times V$.

We shall require that the sets of strategies $\mathscr{S}_P$ and $\mathscr{S}_E$ satisfy Assumption 3.3 and, as in Chapter 3, we shall write

$$\mathscr{V}(x^i, x^s; p, e, \pi^{is}) \triangleq \int_0^{\tau_s} f_0(\mathbf{x}(\tau), \mathbf{u}(\tau), \mathbf{v}(\tau))\, d\tau$$

where $\mathbf{x}$: $\tau \to x = \mathbf{x}(\tau)$, $\tau \in [0, \tau_s]$, represents a path π^{is} in G, $\mathbf{x}(0) = x^i$, $\mathbf{x}(\tau_s) = x^s$, and $\mathbf{u}(\tau) = p(\mathbf{x}(\tau))$, $\mathbf{v}(\tau) = e(\mathbf{x}(\tau))$.

Equations (7.1) and (7.2) constitute a set of $n + 1$ scalar equations which we shall write in vector form

$$d\mathsf{x}/d\tau = \mathsf{f}(\mathsf{x}, u, v) \tag{7.3}$$

where $\mathsf{x} \triangleq (x_0, x_1, \ldots, x_n)$, and

$$\mathsf{f}(\mathsf{x}, u, v) \triangleq (f_0(x, u, v), f_1(x, u, v), \ldots, f_n(x, u, v)).$$

Let $E^{n+1} \triangleq E_n \times \{x_0\}$, $\mathscr{G} \triangleq G \times \{x_0\}$, $\Theta_P \triangleq \theta_P \times \{x_0\}$ and

$$\Theta_P(C) \triangleq \{\mathsf{x}\colon\ \mathsf{x} = (x_0, x), x \in \theta_P, x_0 \leqslant C\} \tag{7.4}$$

where C is a *given* constant.

As the value of parameter C is varied, Eq. (7.4) defines a one-parameter family of targets, namely $\{\Theta_P(C)\}$.

$\mathscr{G}$ and Θ_P are x_0-cylindrical sets whose projections on E^n are G and θ_P, respectively.

Definition 7.2. The set $\mathscr{G}$, the prescribed sets of strategies $\mathscr{S}_P$ and $\mathscr{S}_E$, the state equation (7.3) and

(i) Θ_P in one case;

(ii) $\Theta_P(C)$ in the other case;

and the usual prescriptions for J_P and J_E, define two single-target qualitative games in E^{n+1}. The games corresponding to (i) and (ii) will be termed *Game 2* and *Game 3*, respectively.

Note that the rules are the same for Game 2 and Game 3.

Let x^i be a point of $\mathscr{G}$, and x^i its projection on G. We shall let Π^{is} denote a path in $\mathscr{G}$ from x^i to x^s, and π^{is} a path in G from x^i to x^s.

Let Π^{if} be a path in $\mathscr{G}$ that emanates from x^i, is generated by (p, e), and reaches Θ_P at point x^f. Its projection on G is a path π^{if} that emanates from x^i, is generated by (p, e), and reaches θ_P at x^f, the projection on G of x^f.

Conversely, with each path π^{if} in G that emanates from x^i, is generated by (p, e), and reaches θ_P at point x^f, there is associated a path Π^{if} in $\mathscr{G}$ that emanates from x^i, is generated by (p, e), and reaches Θ_P at point x^f.

As before, let π^{if} be represented by function $\mathbf{x}\colon\ \tau \to x = \mathbf{x}(\tau)$, $\tau \in [0, \tau_f]$; then to π^{if} there corresponds

$$\Pi^{if} = \Big\{\mathsf{x}\colon\ \mathsf{x} = (x_0, x),\quad x = \mathbf{x}(\tau),\qquad x_0 = x_0^{\,i} + \int_0^\tau f_0(\mathbf{x}(\xi), \mathbf{u}(\xi), \mathbf{v}(\xi))\, d\xi,\quad \tau \in [0, \tau_f]\Big\} \tag{7.5}$$

where

$$\mathbf{u}(\tau) = p(\mathbf{x}(\tau)), \qquad \mathbf{v}(\tau) = e(\mathbf{x}(\tau))$$

In other words, π^{if} is the projection on G of Π^{if}.

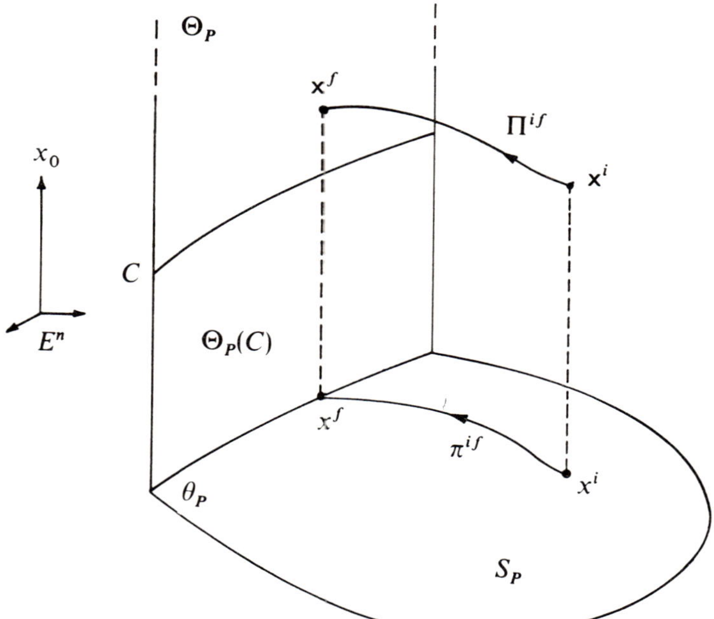

FIG. 7.1. Targets θ_P, Θ_P, and $\Theta_P(C)$; paths Π^{if} and π^{if}.

It follows that, if (p, e) is a playable pair (or a strongly playable pair) for J_P at point $\mathbf{x}^i$ in Game 2, it is a playable pair (or a strongly playable pair) for J_P at point x^i in Game 1, and conversely. We shall let $\mathscr{T}_P(x^i)$ and $\mathscr{T}_P(\mathbf{x}^i)$ denote the sets of playable pairs for J_P at point x^i in Game 1 and at point $\mathbf{x}^i$ in Game 2, respectively. In Game 3, we shall let $\mathscr{T}_P(\mathbf{x}^i;\ C)$ denote the set of playable pairs for J_P at point $\mathbf{x}^i$. The sets of strongly playable pairs will be denoted by $\mathscr{T}_P^*(x^i)$, $\mathscr{T}_P^*(\mathbf{x}^i)$, and $\mathscr{T}_P^*(\mathbf{x}^i; C)$ in Games 1, 2, and 3, respectively.

Likewise, if (p, e) is a playable pair for J_E at point $\mathbf{x}^i$ in Game 2, it is a playable pair for J_E at point x^i in Game 1, and conversely. We shall let $\mathscr{T}_E(x^i)$ and $\mathscr{T}_E(\mathbf{x}^i)$ denote the sets of playable pairs for J_E at point x^i in Game 1 and at point $\mathbf{x}^i$ in Game 2, respectively. We shall let $\mathscr{T}_E(\mathbf{x}^i; C)$ denote the set of playable pairs for J_E at point $\mathbf{x}^i$ in Game 3, and $\mathscr{T}_E^*(x^i)$, $\mathscr{T}_E^*(\mathbf{x}^i)$, $\mathscr{T}_E^*(\mathbf{x}^i; C)$ the corresponding sets of strongly playable pairs in Games 1, 2, and 3, respectively.

7.2 LOCAL SADDLE-POINT CONDITION, GAME 4

If $(p, e) \in \mathscr{T}_P(x^i)$, we shall write

$$V(x^i, \theta_P;\ p, e) \triangleq \int_0^{\tau_f} f_0(\mathbf{x}(\tau), \mathbf{u}(\tau), \mathbf{v}(\tau))\, d\tau \tag{7.6}$$

where $\mathbf{x}$: $\tau \to x = \mathbf{x}(\tau)$, $\tau \in [0, \tau_f]$, represents a path π^{if} that emanates from x^i, is generated by (p, e) and reaches θ_P at x^f. The supporting curve of that path is $\Gamma_{\mathbf{x}}(x^i, p, e)$.

Since, for given x^i and $(p, e) \in \mathscr{T}_P(x^i)$, π^{if} need not be unique, $V(x^i, \theta_P;\ p, e)$ need not be unique. We shall let $V_{\mathbf{x}}(x^i, \theta_P;\ p, e)$ designate the value of $V(x^i, \theta_P;\ p, e)$ defined by function $\mathbf{x}$.

Let x^i be a point of S_P, and let $p^*_{x^i}$ be an optimal strategy for J_P at point x^i in Game 1; that is,

$$(p^*_{x^i}, e) \in \mathscr{T}_P{}^*(x^i) \qquad \forall e \in \mathscr{S}_E \tag{7.7}$$

Then

$$\Gamma_{\mathbf{x}}(x^i, p^*_{x^i}, e) \cap \theta_P \neq \varnothing \qquad \forall \mathbf{x} \in \gamma(x^i, p^*_{x^i}, e), \qquad \forall e \in \mathscr{S}_E$$

and, for $\mathbf{x} \in \gamma(x^i, p^*_{x^i}, e)$, each $\Gamma_{\mathbf{x}}(x^i, p^*_{x^i}, e)$ is the supporting curve of one and only one $\mathscr{P}$-optimal path in G. Consider a member $\pi^{if}_{\mathscr{P}}$ of this set of paths. It follows from Lemma 5.1 that $\pi^{if}_{\mathscr{P}} \subset S_P$. The corresponding path Π^{if} in $S_P \times \{x_0\}$ that emanates from $\mathrm{x}^i = (x_0{}^i, x^i)$ and is given by (7.5) with

$$\mathbf{u}(\tau) = p^*_{x^i}(\mathbf{x}(\tau)), \qquad \mathbf{v}(\tau) = e(\mathbf{x}(\tau))$$

reaches Θ_P at point $\mathrm{x}^f = (x_0{}^f, x^f)$.

By (7.5) and (7.6),

$$x_0{}^f = x_0{}^i + V_{\mathbf{x}}(x^i, \theta_P;\ p^*_{x^i}, e) \tag{7.8}$$

It follows from the definition of $\Theta_P(C)$ that

$$x_0{}^f \leqslant C \Rightarrow \mathrm{x}^f \in \Theta_P(C) \tag{7.9}$$

As a direct consequence of (7.7)–(7.9), the set of points x^i, $x^i \in S_P$, defined by

$$x_0{}^i + V_{\mathbf{x}}(x^i, \theta_P;\ p^*_{x^i}, e) \leqslant C \qquad \forall \mathbf{x} \in \gamma(x^i, p^*_{x^i}, e), \qquad \forall e \in \mathscr{S}_E \tag{7.10}$$

has the following property:

At every point $\mathbf{x}^i$ *of this set*

$$(p^*_{x^i}, e) \in \mathscr{T}_P{}^*(\mathbf{x}^i; C) \qquad \forall e \in \mathscr{S}_E$$

Hence, at every point $\mathbf{x}^i$ of this set, $p^*_{x^i}$ is an optimal strategy for player J_P at that point in Game 3.

Next we shall let $S_P(C)$ denote the set of all points $\mathbf{x}^i$ at which there exists an optimal strategy for player J_P in Game 3. Of course, the set defined by Eq. (7.10) is a subset of $S_P(C)$.

Definition 7.3. We shall say that $\max_{e\in\mathscr{S}_E} V(x^i, \theta_P; p^*_{x^i}, e)$, $x^i \in S_P$, is defined, if there exists M such that

(i) $V_{\mathbf{x}}(x^i, \theta_P; p^*_{x^i}, e) \leqslant M \qquad \forall \mathbf{x} \in \gamma(x^i, p^*_{x^i}, e), \qquad \forall e \in \mathscr{S}_E$

and

(ii) $\exists \hat{e} \in \mathscr{S}_E, \qquad \exists \hat{\mathbf{x}} \in \gamma(x^i, p^*_{x^i}, \hat{e})$ such that $V_{\hat{\mathbf{x}}}(x^i, \theta_P; p^*_{x^i}, \hat{e}) = M$

Then

$$\max_{e\in\mathscr{S}_E} V(x^i, \theta_P; p^*_{x^i}, e) \triangleq M$$

Now we introduce

Assumption 7.1. $\max_{e\in\mathscr{S}_E} V(x^i, \theta_P; p^*_{x^i}, e)$ is defined for all $x^i \in S_P$.

Then the condition

$$x_0{}^i \leqslant C - \max_{e\in\mathscr{S}_E} V(x^i, \theta_P; p^*_{x^i}, e), \qquad x^i \in S_P \tag{7.11}$$

implies (7.10). Hence, the set of points defined by (7.11), namely

$$\Lambda(C) \triangleq \{\mathbf{x}^i : \mathbf{x}^i = (x_0{}^i, x^i),\ x^i \in S_P,\ x_0{}^i \leqslant C - \max_{e\in\mathscr{S}_E} V(x^i, \theta_P; p_x^*, e)\} \tag{7.12}$$

belongs to $S_P(C)$:

$$\Lambda(C) \subseteq S_P(C) \tag{7.13}$$

Next we introduce

Assumption 7.2. There exists $\bar{e}_{x^i} \in \mathscr{S}_E$† such that

$$V(x^i, \theta_P; p, \bar{e}_{x^i}) \geqslant \max_{e\in\mathscr{S}_E} V(x^i, \theta_P; p^*_{x^i}, e) \tag{7.14}$$

for all $x^i \in S_P$, for all $(p, \bar{e}_{x^i}) \in \mathscr{T}_P(x^i)$ and for all values of $V(x^i, \theta_P; p, \bar{e}_{x^i})$.

† Which may depend on x^i.

Let $\varpi(\bar{e}_{x^i})$ be the set of all strategies p such that $(p, \bar{e}_{x^i}) \in \mathscr{T}_P(x^i)$. Clearly, $p^*_{x^i} \in \varpi(\bar{e}_{x^i})$ for all $\bar{e}_{x^i}$.

Definition 7.4. We shall say that $\min_{p\in\varpi(\bar{e}_{x^i})} V(x^i, \theta_P; p, \bar{e}_{x^i})$, $x^i \in S_P$, is defined, if there exists m such that

(i)′ $V(x^i, \theta_P; p, \bar{e}_{x^i}) \geqslant m$

for all $p \in \varpi(\bar{e}_{x^i})$ and for all values of $V(x^i, \theta_P; p, \bar{e}_{x^i})$; and

(ii)′ there exists $\hat{p} \in \varpi(\bar{e}_{x^i})$ and a value of $V(x^i, \theta_P; \hat{p}, \bar{e}_{x^i})$ that is equal to m.

Then $\min_{p\in\varpi(\bar{e}_{x^i})} V(x^i, \theta_P; p, \bar{e}_{x^i}) \triangleq m$.

It follows from (7.14) that condition (i)′ is satisfied at all points $x^i \in S_P$. In view of (7.14) and the definition of $\max_{e\in\mathscr{S}_E} V(x^i, \theta_P; p^*_{x^i}, e)$,

$$V(x^i, \theta_P; p, \bar{e}_{x^i}) \geqslant \max_{e\in\mathscr{S}_E} V(x^i, \theta_P; p^*_{x^i}, e) \geqslant V(x^i, \theta_P; p^*_{x^i}, e)$$

for all $x^i \in S_P$, for all $p \in \varpi(\bar{e}_{x^i})$, for all $e \in \mathscr{S}_E$, and for all values of $V(x^i, \theta_P; p, \bar{e}_{x^i})$ and $V(x^i, \theta_P; p^*_{x^i}, e)$; hence,

$$V(x^i, \theta_P; p^*_{x^i}, \bar{e}_{x^i}) \geqslant \max_{e\in\mathscr{S}_E} V(x^i, \theta_P; p^*_{x^i}, e) \geqslant V(x^i, \theta_P; p^*_{x^i}, \bar{e}_{x^i})$$

for all $x^i \in S_P$ and for all values of $V(x^i, \theta_P; p^*_{x^i}, \bar{e}_{x^i})$.

We conclude that $\min_{p\in\varpi(\bar{e}_{x^i})} V(x^i, \theta_P; p, \bar{e}_{x^i})$ and $V(x^i, \theta_P; p^*_{x^i}, \bar{e}_{x^i})$ are defined for all $x^i \in S_P$.

In other words, we reach the following conclusions:

1. If there is more than one path in S_P starting at point x^i, generated by $(p^*_{x^i}, \bar{e}_{x^i})$, and terminating on θ_P, then all the values of $V(x^i, \theta_P; p^*_{x^i}, \bar{e}_{x^i})$ are equal; and

2. $\min_{p\in\varpi(\bar{e}_{x^i})} V(x^i, \theta_P; p, \bar{e}_{x^i}) = \max_{e\in\mathscr{S}_E} V(x^i, \theta_P; p^*_{x^i}, e) = V(x^i, \theta_P; p^*_{x^i}, \bar{e}_{x^i})$

These results are embodied in

Lemma 7.1. *Let $p^*_{x^i}$ be an optimal strategy for player J_P at point $x^i \in S_P$ in Game 1. If $\max_{e\in\mathscr{S}_E} V(x^i, \theta_P; p^*_{x^i}, e)$ is defined for all $x^i \in S_P$, and if there exists $\bar{e}_{x^i} \in \mathscr{S}_E$ such that*

$$V(x^i, \theta_P; p, \bar{e}_{x^i}) \geqslant \max_{e\in\mathscr{S}_E} V(x^i, \theta_P; p^*_{x^i}, e)$$

for all $x^i \in S_P$, *for all* $(p, \bar{e}_{x^i}) \in \mathscr{T}_P(x^i)$ *and for all values of* $V(x^i, \theta_P;\, p, \bar{e}_{x^i})$, *then*

(i) $V(x^i, \theta_P;\, \overset{*}{p}_{x^i}, \bar{e}_{x^i})$

is defined for all $x^i \in S_P$; *and*

(ii) $V(x^i, \theta_P;\, \overset{*}{p}_{x^i}, e) \leqslant V(x^i, \theta_P;\, \overset{*}{p}_{x^i}, \bar{e}_{x^i}) \leqslant V(x^i, \theta_P;\, p, \bar{e}_{x^i})$

for all $e \in \mathscr{S}_E$, *for all* $p \in \varpi(\bar{e}_{x^i})$, *for all values of* $V(x^i, \theta_P; p^*_{x^i}, e)$ *and* $V(x^i, \theta_P; p, \bar{e}_{x^i})$, *and for all* $x^i \in S_P$; *and*

(iii)

$$\min_{p \in \varpi(\bar{e}_{x^i})} V(x^i, \theta_P;\, p, \bar{e}_{x^i}) = \max_{e \in \mathscr{S}_E} V(x^i, \theta_P;\, \overset{*}{p}_{x^i}, e) = V(x^i, \theta_P;\, \overset{*}{p}_{x^i}, \bar{e}_{x^i})$$

Condition (ii) is a *local saddle-point condition.*

Now, if there exists $p^* \in \mathscr{S}_P$ that is optimal on S_P in Game 1, if $\max_{e \in \mathscr{S}_E} V(x^i, \theta_P; p^*, e)$ is defined for all $x^i \in S_P$, and if there exists $\bar{e} \in \mathscr{S}_E$ such that

$$V(x^i, \theta_P;\, p, \bar{e}) \geqslant \max_{e \in \mathscr{S}_E} V(x^i, \theta_P;\, p^*, e)$$

for all $(p, \bar{e}) \in \mathscr{T}_P(x^i)$, for all values of $V(x^i, \theta_P; p, \bar{e})$ and for all $x^i \in S_P$, then (ii) is identical with the saddle-point condition given in Chapter 2; namely, it becomes

$$V(x^i, \theta_P; p^*, e) \leqslant V(x^i, \theta_P; p^*, \bar{e}) \leqslant V(x^i, \theta_P; p, \bar{e})$$

for all $x^i \in S_P$, for all $(p, \bar{e}) \in \mathscr{T}_P(x^i)$ and for all values of $V(x^i, \theta_P; p^*, e)$ and $V(x^i, \theta_P; p, \bar{e})$.

From the point of view of quantitative games,

$$V(x^i, \theta_P;\; p, e), \qquad (p, e) \in \mathscr{T}_P(x^i),$$

is the *cost functional* associated with a terminating path in S_P that emanates from x^i, is generated by strategy pair (p, e), and reaches target θ_P.

Definition 7.5. The quantitative zero-sum, two-player game which has the same rules as qualitative Game 1, and whose objectives are defined in terms of θ_P† and $V(x^i, \theta_P; p, e)$, will be called *Game 4*.

† That is, $\theta = \theta_P$.

In Game 4, p^* and $\bar{e}$ are optimal strategies on S_P for players J_P and J_E, respectively.

7.3 SURFACES OF THE GAME

Let us consider points $\mathbf{x}^i = (x_0{}^i, x^i)$ such that

$$x_0{}^i = C - \max_{e \in \mathscr{S}_E} V(x^i, \theta_P; \overset{*}{p}_{x^i}, e), \quad x^i \in S_P$$

or equivalently

$$x_0{}^i + V(x^i, \theta_P; \overset{*}{p}_{x^i}, \bar{e}_{x^i}) = C, \quad x^i \in S_P \tag{7.15}$$

Definition 7.6. Equation (7.15) is the equation of a set of points $\mathbf{x}^i$ in $S_P \times \{x_0\}$; that is, (7.15) defines a surface which we shall call a *generalized surface of the game* $\hat{\Sigma}(C)$ *in Game 4.*

This surface has properties which are similar to those of a surface of the game $\Sigma(C)$.

When functions $\overset{*}{p}_{x^i}$ and $\bar{e}_{x^i}$ do not depend on $x^i \in S_P$, then $\hat{\Sigma}(C)$ and $\Sigma(C)$ coincide for all $x^i \in S_P$.

Surface $\hat{\Sigma}(C)$ depends on parameter C. Variable $x_0{}^i$ is the *cost variable.* Since $V(x^i, \theta_P; \overset{*}{p}_{x^i}, \bar{e}_{x^i})$ does not contain the cost variable, Eq. (7.15) defines a one parameter family of surfaces, $\{\hat{\Sigma}(C)\}$, such that any two surfaces of the family can be deduced from one another by a translation parallel to the cost axis.

Now consider a path Π^{if} that emanates from $\mathbf{x}^i \in S_P \times \{x_0\}$, is generated by $(p, \bar{e}_{x^i})$ where $p \in \varpi(\bar{e}_{x^i})$, and reaches Θ_P at point $\mathbf{x}^f = (x_0{}^f, x^f)$. We have

$$\begin{aligned} x_0{}^f &= x_0{}^i + V(x^i, \theta_P; p, \bar{e}_{x^i}) \\ &\geqslant x_0{}^i + \max_{e \in \mathscr{S}_E} V(x^i, \theta_P; \overset{*}{p}_{x^i}, e) \end{aligned} \tag{7.16}$$

for all $p \in \varpi(\bar{e}_{x^i})$ and for all values of $V(x^i, \theta_P; p, \bar{e}_{x^i})$.

From (7.16) and from the fact that

$$V(x^i, \theta_P; \overset{*}{p}_{x^i}, \bar{e}_{x^i}) = \max_{e \in \mathscr{S}_E} V(x^i, \theta_P; \overset{*}{p}_{x^i}, e),$$

we deduce that

$$\begin{aligned} &x_0{}^i > C - V(x^i, \theta_P; \overset{*}{p}_{x^i}, \bar{e}_{x^i}) \\ &\quad\Rightarrow x_0{}^i + V(x^i \theta_P; p, \bar{e}_{x^i}) > C \end{aligned} \tag{7.17}$$

for all $p \in \varpi(\bar{e}_{x^i})$ and for all values of $V(x^i, \theta_P; p, \bar{e}_{x^i})$.

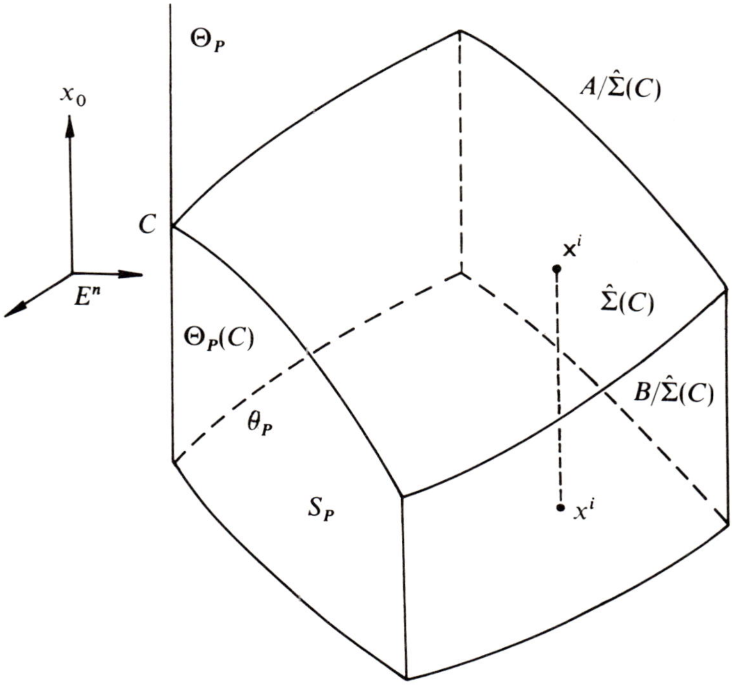

FIG. 7.2. Generalized surface of the game in Game 4.

Hence, at every point $\mathbf{x}^i$ of the set $A/\hat{\Sigma}(C)$ defined by

$$A/\hat{\Sigma}(C) \triangleq \{\mathbf{x}^i\colon \quad \mathbf{x}^i = (x_0{}^i, x^i), \quad x^i \in S_P, \\ x_0{}^i > C - V(x^i, \theta_P; \overset{*}{p}_{x^i}, \bar{e}_{x^i})\} \quad (7.18)$$

there exists a strategy e, namely $\bar{e}_{x^i}$, such that whatever the strategy p, $(p, \bar{e}_{x^i})$ is a playable pair for J_E at that point in Game 3; that is,

$$(p, \bar{e}_{x^i}) \in \mathscr{T}_E(\mathbf{x}^i; C) \qquad \forall p \in \mathscr{S}_P, \quad \forall \mathbf{x}^i \in A/\hat{\Sigma}(C) \qquad (7.19)$$

Since player J_E has no target of his own, it follows from the definition of a playable pair that relation (7.19) coincides with

$$(p, \bar{e}_{x^i}) \in \mathscr{T}_E{}^*(\mathbf{x}^i; C) \qquad \forall p \in \mathscr{S}_P, \quad \forall \mathbf{x}^i \in A/\hat{\Sigma}(C) \qquad (7.20)$$

Equation (7.19), and hence Eq. (7.20), is a consequence of (7.17) if $p \in \varpi(\bar{e}_{x^i})$, and it is trivial if $p \notin \varpi(\bar{e}_{x^i})$.

Next we shall let $S_E(C)$ denote the set of all points $\mathbf{x}^i$ at which there exists an optimal strategy for player J_E in Game 3.

It follows from (7.20) that

$$A/\hat{\Sigma}(C) \subset S_E(C) \tag{7.21}$$

Similarly, consider

$$B/\hat{\Sigma}(C) \triangleq \{\mathbf{x}^i: \quad \mathbf{x}^i = (x_0{}^i, x^i). \quad x^i \in S_P, \quad x_0{}^i < C - V(x^i, \theta_P;\, p^*_{x^i}, \bar{e}_{x^i})\} \tag{7.22}$$

Since $V(x^i, \theta_P; p^*_{x^i}, \bar{e}_{x^i}) = \max_{e \in \mathscr{S}_E} V(x^i, \theta_P; p^*_{x^i}, e)$,

$$B/\hat{\Sigma}(C) \subset \Lambda(C)$$

where $\Lambda(C)$ is the set defined by (7.12). Then it follows from (7.13) that

$$B/\hat{\Sigma}(C) \subset S_P(C) \tag{7.23}$$

Furthermore,

$$\widetilde{A/\hat{\Sigma}}(C) \triangleq (A/\hat{\Sigma}(C)) \cup \hat{\Sigma}(C) \subseteq \overline{A/\hat{\Sigma}(C)} \subseteq \overline{S_E(C)}$$

$$\widetilde{B/\hat{\Sigma}}(C) \triangleq (B/\hat{\Sigma}(C)) \cup \hat{\Sigma}(C) \subseteq \overline{B/\hat{\Sigma}(C)} \subseteq \overline{S_P(C)}$$

Hence,

$$\hat{\Sigma}(C) = \widetilde{A/\hat{\Sigma}}(C) \cap \widetilde{B/\hat{\Sigma}}(C) \subseteq \overline{S_E(C)} \cap \overline{S_P(C)} \tag{7.24}$$

But $\Xi(C) = \overline{S_P(C)} \cap \overline{S_E(C)}$ is the surface of the game in qualitative Game 3. Consequently,

$$\hat{\Sigma}(C) \subseteq \Xi(C) \tag{7.25}$$

7.4 A CONNECTION BETWEEN PATHS IN GAMES 1 AND 4

Now we introduce

Assumption 7.3. Strategies $p^*_{x^i}$ and $\bar{e}_{x^i}$ do not depend on $x^i \in S_P$. We shall denote these strategies by p^* and $\bar{e}$.

Then, as pointed out before,

(i) $(p^*, \bar{e})$ is an optimal strategy pair on S_P *in Game 4;* and

(ii) $\hat{\Sigma}(C)$ and $\Sigma(C)$ coincide for all $x^i \in S_P$.

Let $\pi_{\mathscr{P}}$ be a $\mathscr{P}$-optimal path *in Game 1*, emanating from $x^i \in S_P$ and generated by strategy pair $(p^*, \bar{e})$. By Lemma 5.1,

$$\pi_{\mathscr{P}} \subset S_P$$

Let $\Pi^*(C)$ be an optimal path *in Game 4*, generated by strategy pair $(p^*, \bar{e})$, whose projection on G is $\pi^* = \pi_{\mathscr{P}}$. From Corollary 2.1, with $\mathscr{X}^* = S_P \times \{x_0\}$, $\Pi^*(C) \subset \Sigma(C)$.

And so, finally, we arrive at

Theorem 7.1. *If $p^*_{x^i}$ is an optimal strategy for player J_P at point $x^i \in S_P$ in Game 1; if there exist a cost $V(x^i, \theta_P; p^*_{x^i}, e)$ satisfying Assumption 7.1 and a strategy $\bar{e}_{x^i} \in \mathscr{S}_E$ satisfying Assumption 7.2; if Assumption 7.3 is satisfied, that is, if $p^*_{x^i} = p^*$ and $\bar{e}_{x^i} = \bar{e}$ do not depend on x^i; then every $\mathscr{P}$-optimal path $\pi_{\mathscr{P}}$ in Game 1, emanating from x^i and generated by strategy pair $(p^*, \bar{e})$, is the projection on S_P of an optimal path $\Pi^*(C)$ in Game 4, generated by strategy pair (p^*, e^*), $e^* = \bar{e}$.*

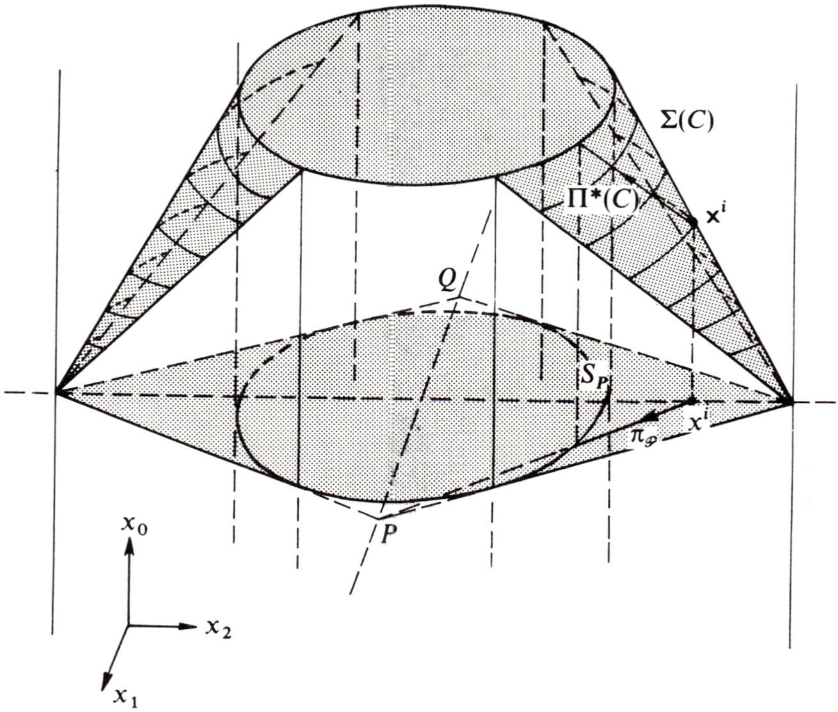

FIG. 7.3. Connection between paths in Games 1 and 4. Game 1 is Example 6.2; Game 4 is Example 3.1; $\Sigma(C)$ is game surface in the x_0–x_1–x_2 space, for Example 3.1. Sections of $\Sigma(C)$ by planes parallel to x_1–x_2 are shown on this figure.

Of course, there may be more than one quantitative Game 4 associated with a given qualitative Game 1.

Theorem 7.1 is illustrated by Examples 6.2 and 3.1. We shall let the reader verify that if Example 6.2 is Game 1, then Example 3.1 is a Game 4 associated with it. The corresponding game surface $\Sigma(C)$ is pictured on Fig. 7.3 in the x_0–x_1–x_2 space. This figure should be compared to Figs. 3.9 and 6.7.

Appendix

Let G be a domain in E^n; that is, an open connected set in the topology of E^n.

The topology induced by E^n on G (i.t.) is defined by the following

Axiom. *An open set in the i.t. is a subset of G which is open in E^n.*

A subset of G is closed in the i.t. if its complement in G is open. It follows that

1′. G is closed in the i.t.

2′. $\varnothing$ is closed in the i.t.

Let S be a subset of G. Point x is an interior point of S in the i.t. if $x \in S$ and there exists an open set in the i.t. which contains x. The interior of S in the i.t. is

$$\mathring{S} \triangleq \{x: \quad x \text{ is an interior point of } S\}$$

It follows that

$$\mathring{S} \subseteq S$$

Let X be a subset of G, and let

$$Y = \text{comp } X$$

where comp X means the complement of X in G. The closure of X in the i.t. is

$$\bar{X} \triangleq \text{comp } \mathring{Y}$$

It follows that

$$X \subseteq \bar{X}$$

$$\text{comp } \bar{X} = \overset{\circ}{\widehat{\text{comp } X}}$$

$$\overline{\text{comp } Y} = \text{comp } \mathring{Y}$$

The boundary of X in the i.t. is

$$\partial X \triangleq \bar{X} \cap \overline{\text{comp } X}$$

We have

$$\partial X = \bar{X} \cap \text{comp } \mathring{X}$$

$$\begin{aligned} \mathring{X} \cup \partial X &= \mathring{X} \cup (\bar{X} \cap \text{comp } \mathring{X}) \\ &= (\mathring{X} \cup \bar{X}) \cap (\mathring{X} \cup \text{comp } \mathring{X}) \\ &= \bar{X} \cap (\mathring{X} \cup \text{comp } \mathring{X}) \end{aligned}$$

It follows that

$$\begin{aligned} \text{comp } (\mathring{X} \cup \partial X) &= \text{comp } \bar{X} \cup (\text{comp } \mathring{X} \cap \text{comp comp } \mathring{X}) \\ &= \text{comp } \bar{X} \cup (\mathring{X} \cap \text{comp } \mathring{X}) \end{aligned}$$

Furthermore, we have

$$\mathring{X} \cap \text{comp } \mathring{X} = \varnothing$$

Hence

$$\text{comp } (\mathring{X} \cup \partial X) = \text{comp } \bar{X}$$

Hence

$$\bar{X} = \mathring{X} \cup \partial X$$

REFERENCES

On the Geometry of Optimal Processes

1. A. Blaquiere and G. Leitmann, On the Geometry of Optimal Processes, Parts I, II, III, Univ. California, Berkeley, IER Repts. AM-64-10, AM-65-11, AM-66-1.
2. G. Leitmann, Some Geometrical Aspects of Optimal Processes, *SIAM J. Control* **3,** 53 (1965).
3. A. Blaquiere, Further Investigation into the Geometry of Optimal Processes, *SIAM J. Control* **3,** 19 (1965).
4. A. Blaquiere and G. Leitmann, Some Geometric Aspects of Optimal Processes, Part I: Problems with Control Constraints, *Proc. Congr. Automatique Théorique, Paris 1965.* Dunod, Paris, 1967.
5. K. V. Saunders and G. Leitmann, Some Geometric Aspects of Optimal Processes, Part II: Problems with State Constraints, *Proc. Congr. Automatique Théorique, Paris 1965.* Dunod, Paris, 1967.
6. A. Blaquiere and G. Leitmann, On the Geometry of Optimal Processes, *in* "Topics in Optimization" (G. Leitmann, ed.), pp. 265–371. Academic Press, New York, 1967.
7. H. Halkin, The Principle of Optimal Evolution, *in* "Nonlinear Differential Equations and Nonlinear Mechanics" (J. P. LaSalle and S. Lefschetz, eds.). Academic Press, New York, 1963.
8. H. Halkin, Mathematical Foundations of System Optimization, *in* "Topics in Optimization" (G. Leitmann, ed.), pp. 198–260. Academic Press, New York, 1967.
9. E. Roxin, A Geometric Interpretation of Pontryagin's Maximum Principle, *in* "Nonlinear Differential Equations and Nonlinear Mechanics" (J. P. LaSalle and S. Lefschetz, eds.). Academic Press, New York, 1963.
10. R. E. Bellman, "Dynamic Programming." Princeton Univ. Press, Princeton, New Jersey, 1957.
11. G. Leitmann, "An Introduction to Optimal Control." McGraw-Hill, New York, 1966.
12. A. Blaquiere and G. Leitmann, Further Geometric Aspects of Optimal Processes: Multiple-Stage Dynamic Systems, *in* "Mathematical Theory of Control" (A. V. Balakrishnan and L. W. Neustadt, eds.). Academic Press, New York, 1967.

On The Theory of Games

13. J. von Neumann and O. Morgenstern, "Theory of Games and Economic Behavior." Princeton Univ. Press, Princeton, New Jersey, 1953.
14. J. McKinsey, "An Introduction to the Theory of Games." McGraw-Hill, New York, 1952.
15. L. Zadeh, Optimality and Nonscalar-valued Performance Criteria, *IEEE Trans. Automatic Control* **AC-8,** 59 (1963).
16. R. Isaacs, "Differential Games." Wiley, New York, 1965.
17. L. D. Berkovitz, A Variational Approach to Differential Games, *in* "Advances in Game Theory," pp. 127–174. Princeton Univ. Press, Princeton, New Jersey, 1964.

18. L. D. BERKOVITZ AND W. H. FLEMING, On Differential Games with Integral Payoff, *in* "Contributions to the Theory of Games III," pp. 413–435. Princeton Univ. Press, Princeton, New Jersey, 1957.
19. L. D. BERKOVITZ, Necessary Conditions for Optimal Strategies in a Class of Differential Games and Control Problems, *SIAM J. Control* **5,** 1 (1967).
20. D. L. KELENDZHERIDZE, A Pursuit Problem, *in* "The Mathematical Theory of Optimal Processes." Wiley (Interscience), New York, 1962.
21. L. S. PONTRYAGIN, On Some Differential Games, *SIAM J. Control* **3,** 49 (1965).
22. L. S. PONTRYAGIN, On the Theory of Differential Games, *Uspehi Mat. Nauk* **21,** 219 (1966).
23. E. F. MISHCHENKO AND L. S. PONTRYAGIN, Linear Differential Games, *Dokl. Akad. Nauk SSSR* **174,** 27 (1967) [English transl.]: *Soviet Math. Dokl.* **8,** 585 (1967).
24. Y. C. HO, A. E. BRYSON, AND S. BARON, Differential Games and Optimal Pursuit-Evasion Strategies, *IEEE Trans. Automatic Control* **AC-10,** 385 (1965).
25. S. BARON, Differential Games and Optimal Pursuit-Evasion Strategies. Doctoral dissertation in Applied Mathematics, Harvard University, Cambridge, Massachusetts 1966.
26. E. N. SIMAKOVA, Differential Games (a survey paper), *Avtomat. i Telemeh.* **27,** 161 (1966).
27. I. G. SARMA AND R. K. RAGADE, Some Considerations in Formulating Optimal Control Problems as Differential Games, *Internat. J. Control* **4,** 265 (1966).
28. A. KAUFMANN, "Graphs, Dynamic Programming, and Finite Games." Academic Press, New York, 1967.
29. G. LEITMANN AND G. MON, Some Geometric Aspects of Differential Games, *J. Astronaut. Sci.* **14,** 56 (1967).
30. G. LEITMANN AND G. MON, On a Class of Differential Games, *in* "Proceedings of the Colloq. Advanced Problems and Methods for Space Flight Optimization, University of Liége, 1967" (B. Fraeijs de Veubeke, ed.) Pergamon Press, Oxford, 1968.
31. G. MON, Differential Game Theory: A Geometric Approach. Unpublished doctoral dissertation, Univ. California, Berkeley, 1967.
32. A. BLAQUIERE AND G. LEITMANN, "Jeux Quantitatifs, Mémorial des Sciences Mathématiques." Gauthier-Villars, Paris, 1969.
33. G. WANG AND G. LEITMANN, Necessary and sufficient Conditions for Two-Person, Zero-Sum Multistage Games, *J. Optim. Theory Appl.* (1969, in press).
34. A. BLAQUIERE AND G. LEITMANN, Multistage Quantitative Games, *in* "Proceedings of 2nd Hawaii International Conference on System Sciences, 1969," p. 579. Western Periodicals Co., North Hollywood, California.
35. A. BLAQUIERE AND N. GANI, Multistage Quantitative Games with Unprescribed Terminal Stage, *Comp. Rend. Acad. Sci. Paris,* **268,** 428 (1969).
36. A. BLAQUIERE AND G. LEITMANN, Multistage Quantitative Games with Unprescribed Terminal Stage, *Internat. J. Non-Linear Mech.* (1969, in press).

On Qualitative Games

37. A. BLAQUIERE AND F. GERARD, "On the Geometry of Optimal Strategies in Two-Person Games of Kind." Mehanika, Moscow, 1968.
38. A. BLAQUIERE AND F. GERARD, On the Geometry of Optimal Strategies in Two-Person Games of Kind, *J. Computer System Sci.* Vol. 2, No 3, October 1968.

39. F. GERARD, Théorie géométrique des jeux différentiels qualitatifs à deux joueurs. Doctoral dissertation, Faculté des Sciences de Paris, 1968.

On Differential Equations and System Theory

40. E. A. CODDINGTON AND N. LEVINSON, "Theory of Ordinary Differential Equations." McGraw-Hill, New York, 1955.
41. G. SANSONE AND R. CONTI, "Non-Linear Differential Equations." Pergamon Press, Oxford, 1964.
42. W. A. PORTER, "Modern Foundations of Systems Engineering." Macmillan, New York, 1966.

Subject Index

List of Assumptions

List of Corollaries

List of Lemmas

List of Theorems

Mathematics in Science and Engineering

A Series of Monographs and Textbooks

Edited by RICHARD BELLMAN, *University of Southern California*

1. T. Y. Thomas. Concepts from Tensor Analysis and Differential Geometry. Second Edition. 1965

2. T. Y. Thomas. Plastic Flow and Fracture in Solids. 1961

3. R. Aris. The Optimal Design of Chemical Reactors: A Study in Dynamic Programming. 1961

4. J. LaSalle and S. Lefschetz. Stability by by Liapunov's Direct Method with Applications. 1961

5. G. Leitmann (ed.). Optimization Techniques: With Applications to Aerospace Systems. 1962

6. R. Bellman and K. L. Cooke. Differential-Difference Equations. 1963

7. F. A. Haight. Mathematical Theories of Traffic Flow. 1963

8. F. V. Atkinson. Discrete and Continuous Boundary Problems. 1964

9. A. Jeffrey and T. Taniuti. Non-Linear Wave Propagation: With Applications to Physics and Magnetohydrodynamics. 1964

10. J. T. Tou. Optimum Design of Digital Control Systems. 1963.

11. H. Flanders. Differential Forms: With Applications to the Physical Sciences. 1963

12. S. M. Roberts. Dynamic Programming in Chemical Engineering and Process Control. 1964

13. S. Lefschetz. Stability of Nonlinear Control Systems. 1965

14. D. N. Chorafas. Systems and Simulation. 1965

15. A. A. Pervozvanskii. Random Processes in Nonlinear Control Systems. 1965

16. M. C. Pease, III. Methods of Matrix Algebra. 1965

17. V. E. Benes. Mathematical Theory of Connecting Networks and Telephone Traffic. 1965

18. W. F. Ames. Nonlinear Partial Differential Equations in Engineering. 1965

19. J. Aczel. Lectures on Functional Equations and Their Applications. 1966

20. R. E. Murphy. Adaptive Processes in Economic Systems. 1965

21. S. E. Dreyfus. Dynamic Programming and the Calculus of Variations. 1965

22. A. A. Fel'dbaum. Optimal Control Systems. 1965

23. A. Halanay. Differential Equations: Stability, Oscillations, Time Lags. 1966

24. M. N. Oguztoreli. Time-Lag Control Systems. 1966

25. D. Sworder. Optimal Adaptive Control Systems. 1966

26. M. Ash. Optimal Shutdown Control of Nuclear Reactors. 1966

27. D. N. Chorafas. Control System Functions and Programming Approaches (In Two Volumes). 1966

28. N. P. Erugin. Linear Systems of Ordinary Differential Equations. 1966

29. S. Marcus. Algebraic Linguistics; Analytical Models. 1967

30. A. M. Liapunov. Stability of Motion. 1966

31. G. Leitmann (ed.). Topics in Optimization. 1967

32. M. Aoki. Optimization of Stochastic Systems. 1967

33. H. J. Kushner. Stochastic Stability and control. 1967

34. M. Urabe. Nonlinear Autonomous Oscillations. 1967

35. F. Calogero. Variable Phase Approach to Potential Scattering. 1967

36. A. Kaufmann. Graphs, Dynamic Programming, and Finite Games. 1967

37. A. Kaufmann and R. Cruon. Dynamic Programming: Sequential Scientific Management. 1967

38. J. H. Ahlberg, E. N. Nilson, and J. L. Walsh. The Theory of Splines and Their Applications. 1967

39. Y. Sawaragi, Y. Sunahara, and T. Nakamizo. Statistical Decision Theory in Adaptive Control Systems. 1967

40. R. Bellman. Introduction to the Mathematical Theory of Control Processes Volume I. 1967 (Volumes II and III in preparation)

41. E. S. Lee. Quasilinearization and Invariant Imbedding. 1968

42. W. Ames. Nonlinear Ordinary Differential Equations in Transport Processes. 1968

43. W. Miller, Jr. Lie Theory and Special Functions. 1968

44. P. B. Bailey, L. F. Shampine, and P. E. Waltman. Nonlinear Two Point Boundary Value Problems. 1968.

45. Iu. P. Petrov. Variational Methods in Optimum Control Theory. 1968

46. O. A. Ladyzhenskaya and N. N. Ural't-seva. Linear and Quasilinear Elliptic Equations. 1968

47. A. Kaufmann and R. Faure. Introduction to Operations Research. 1968

48. C. A. Swanson. Comparison and Oscillation Theory of Linear Differential Equations. 1968

49. R. Hermann. Differential Geometry and the Calculus of Variations. 1968

50. N. K. Jaiswal. Priority Queues. 1968

51. H. Nikaido. Convex Structures and Economic Theory. 1968

52. K. S. Fu. Sequential Methods in Pattern Recognition and Machine Learning. 1968

53. Y. L. Luke. The Special Functions and Their Approximations (In Two Volumes). 1969

54. R. P. Gilbert. Function Theoretic Methods in Partial Differential Equations. 1969

55. V. Lakshmikantham and S. Leela. Differential and Integral Inequalities (In Two Volumes.) 1969

56. S. H. Hermes and J. P. LaSalle. Functional Analysis and Time Optimal Control. 1969.

57. M. Iri. Network Flow, Transportation, and Scheduling: Theory and Algorithms. 1969

58. A. Blaquiere, F. Gerard, and G. Leitmann. Quantitative and Qualitative Games. 1969

59. P. Falb and J. De Jong. Successive Approximation Methods in Control and Oscillation Theory. 1969

In preparation

R. Bellman. Methods of Nonlinear Analysis, Volume I

R. Bellman, K. L. Cooke, and J. A. Lockett. Algorithms, Graphs, and Computers

A. H. Jazwinski. Stochastic Processes and Filtering Theory

S. R. McReynolds and P. Dyer. The Computation and Theory of Optimal Control

J. M. Mendel and K. S. Fu. Adaptive, Learning, and Pattern Recognition Systems: Theory and Applications

G. Rosen. Formulations of Classical and Quantum Dynamical Theory

E. J. Beltrami. Methods of Nonlinear Analysis and Optimization

H. H. Happ. The Theory of Network Diakoptics